Sprir

AF760535

CISM COURSES AND LECTURES

The series presents lecture notes, monographs, edited works and proceedings in the field of Mechanics, Engineering, Computer Science and Applied Mathematics.
Purpose of the series is to make known in the international scientific and technical community results obtained in some of the activities organized by CISM, the International Centre for Mechanical Sciences.

INTERNATIONAL CENTRE FOR MECHANICAL SCIENCES

COURSES AND LECTURES - No. 494

ADVANCED EARTHQUAKE ENGINEERING ANALYSIS

EDITED BY

ALAIN PECKER
ECOLE POLYTECHNIQUE, PALAISEAU, FRANCE

SpringerWienNewYork

This volume contains 122 illustrations

SPIN 12048397

All contributions have been typeset by the authors.

ISBN 978-3-211-74213-6 SpringerWienNewYork

PREFACE

This book contains lecture notes, albeit not covering all materials, delivered at the course on Advanced Earthquake Engineering Analysis that took place at CISM Udine in July 3-7, 2005.

During the last decade, the state of the Art in Earthquake Engineering Design has made significant steps towards a more rationale analysis of structures. Scientists have long recognized that earthquake design is guided by displacements and deformations rather than forces. However due to the historical background of structural engineers in static analyses, effects of earthquake on structures have been viewed as forces acting on the structures. All presently available design building codes are developed along these lines.

Our knowledge of ground motion characteristics, earthquake geotechnical engineering, structural behavior (design and numerical analyses) and model tests have advanced to a point where it is possible to anticipate a significant move from force based design approaches to displacements based design approaches. Although displacement based design approaches constitute the kernel of most research programs, they have not yet been incorporated in the State of Practice.

The purpose of the course was to review the fundamentals of displacement based methods, starting from engineering seismology, earthquake geotechnical engineering, to focus on design, analysis and testing of structures with emphasis on buildings and bridges.

The five main topics presented during the course are detailed below. Each topic started with the fundamentals and then focused on advanced, State of the Art, subjects. The lectures were heavily illustrated with examples drawn from actual projects in which the lecturers are deeply involved.

- *Engineering Seismology: measurement and characterisation of seismic motions; prediction of ground motions with empirical attenuation relationships; physics behind earthquake signals; site effects.*
- *Geotechnical Engineering : non linear soil behavior under cyclic loading ; non linear site response analyses; soil-structure interaction including non linear effects; earthquake resistant design of foundation; performance based design.*
- *Seismic Analyses: introduction to structural response and computer modelling; overview of linear elastic analysis methods; non linear inelastic analyses; pushover analyses; non linear time history analysis.*
- *Seismic Design: need for displacement based design; fundamentals for direct displacement based design and assessment; strength and deformation capacity.*

- *Seismic Testing: introduction to seismic testing; shake table tests; pseudo dynamic tests; centrifuge testing; special topics and illustration.*

The coordinator of the course wishes to express his gratitude to P.Y. Bard, G.M Calvi, R. Pinho, N.Priestley and P. Sollogoub for their active participation in the elaboration, preparation and delivery of the course. He and the lecturers are also indebted to Pr. Jean Salençon who, as non resident Rector at CISM, suggested the topic of the course and provided several constructive suggestions during its preparation. They also want to express their gratitude to the Secretariat staff of CISM for their efficient handling of administrative matters before and during the course.

CONTENTS

Soil Behaviour under Cyclic Loading

Alain Pecker[1,2]

[1] Solid Mechanics Laboratory, CNRS UMR 7649, Department of Mechanics, Ecole Polytechnique, Palaiseau, France
[2] Géodynamique et Structure, Bagneux, France

Abstract. Fundamental characteristics of soil behaviour during earthquakes are reviewed; field and laboratory evidences of non linearities and energy dissipation mechanisms are presented. Different kinds of soil constitutive models are discussed with special emphasis on the equivalent linear viscoelastic model commonly used in engineering practice.

1 Field Observation of Soil Behaviour during Earthquakes

Field observations made on sites during earthquakes clearly point out the influence of the geotechnical nature of the soil profile on the recorded ground surface motion. This fact has been acknowledged more than twenty years ago with the pioneering work of Seed *et al* (1976) and lead to the consideration, in design practice and in seismic building code, of different spectral shapes to specify the seismic action.

All recent major earthquakes recorded worldwide (Mexico, 1985; Loma Prieta, 1989; Northridge, 1994; Kobe, 1995) have all confirmed the following observation: in general, soft alluvial deposits tend to amplify the incident ground motion, especially in the low frequency range. To illustrate that proposal, let us consider the records from the Loma Prieta earthquake; on rock sites, in the town of San Francisco, all records exhibit a peak acceleration of the order of 0.1g; records from the same earthquake, at the surface of alluvial sites, exhibit peak accelerations two to three times larger (Table 1, AFPS, 1990) and acceleration response spectra with significant amplification in the low frequency range. Since the epicentre of the earthquake is remote from the recording stations (approximately 85 km), this amplification cannot be attributed to a source directivity effect; accordingly, all the recording sites being close to one another (within a radius of a few kilometres), it cannot either be attributed to a path affect between the focus and the site. The only parameter that can affect the ground recorded motion is the geotechnical nature of the soil profile, i.e. the mechanical characteristics of the soil layers close to the ground surface.

However, the previously reported observations, and several others like in Mexico City, must not be misleading: alluvial deposits do not invariably amplify the incident ground motion. Looking again at observations made in San Francisco, but during the 1957 earthquake, it appears that although the recorded accelerations on the rock sites were again of the order of 0.1g (because of the smaller magnitude but closer distance to the town), the recorded accelerations on the soil sites were between 1.5 and 2 times smaller than on the rock (Table 1).

Table 1. Recorded peak ground acceleration in San Francisco

Recording station	Soil profile	Peak ground surface acceleration (g)	
		1957	1989
Golden Gate Park	Rock	0.13	
Market/Guerrero St	Rock	0.12	
State Building	Sand + clayey sand (60 m)	0.10	
Mason/Pine St	Rock	0.10	
Alexander Building	Clayey silt + Sand (45 m)	0.07	0.17
Southern Pacific B.	Soft clay	0.05	0.20
Rincon Hill	Rock	0.10	0.09
Oakland City Hall	Clay, Sand (30 m) + Stiff clay (270 m)	0.04	0.26

The previous observations clearly show that the response of a soil deposit depends on the frequency content of the incident motion: the 1957 event has a smaller magnitude and a closer distance to the recording sites; its frequency content is richer in high frequencies than the larger, far away event. Another factor, which has a significant impact on the ground response, is the level of shaking induced by the earthquake; the higher this level, the larger the strains induced in the ground. These large strains induce a non linear response of the soil as shown in Figure 1. This figure depicts the Fourier amplitude spectra of the main shock (solid line) and of the aftershocks (shaded area), scaled by the same quantity at a rock outcrop, of Loma Prieta earthquake records at Treasure Island. Obviously would the soil profile behave linearly, the amplification with respect to the rock outcrop would be independent of the generating event. In this case smaller amplifications occur for the strongest event, which is consistent with the larger strains induced in the profile creating softening of the soil deposit.

In order to be able to predict such phenomena a deep understanding of the soil behaviour under cyclic loading is mandatory. It can be stated that nowadays, although many aspects still remain to be clarified, our knowledge of soil behaviour has advanced to a point where constitutive modelling can be reliably employed to allow for accurate prediction in engineering practice.

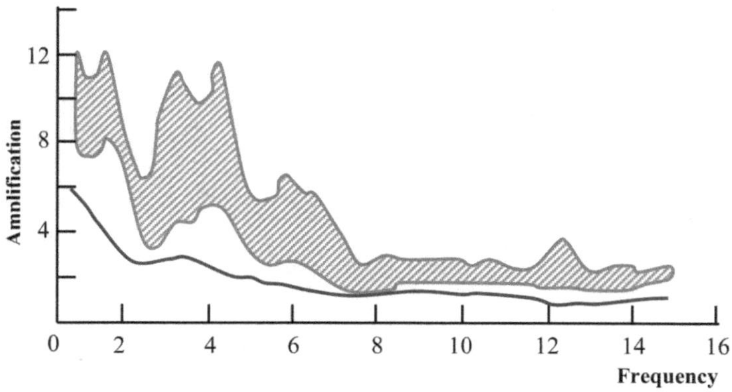

Figure 1. Fourier amplitude spectra of the main shock (solid line) and of the aftershocks (shaded area) of the Loma Prieta records at Treasure Island (after Jarpe *et al*, 1989)

2 Experimental Description of Soil Behaviour

It is commonly admitted for site response analyses, or for soil structure interaction problems, to consider that the seismic horizontal motion is caused by the vertical propagation of horizontally polarized shear waves. Under those conditions, a soil element within the soil profile is subjected to stress cycles similar to those presented in Figure 2.

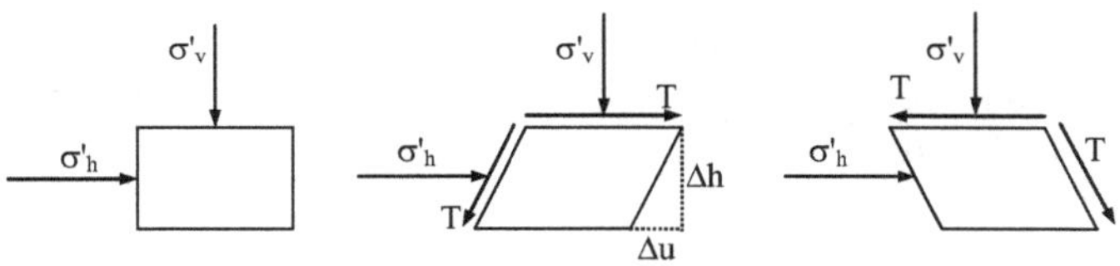

Figure 2. Idealized stress cycle during an earthquake

Initially, for a horizontally layered profile, the soil element is in equilibrium under the vertical effective stress σ'_v and the horizontal effective stress $K_0 \sigma'_v$ where K_0 is the at rest earth pressure coefficient. When the wave travels trough the soil profile an additional shear stress $\tau(t)$ is superimposed on the horizontal faces of the soil element and, hence on the vertical ones to maintain equilibrium conditions. Under the action of this stress the soil element undergoes a shear strain, which for an elastic material is accompanied by a zero volumetric strain. The shear strain, also called distortion, is defined by (eq.(1)):

$$\gamma = \frac{\Delta u}{\Delta h} \tag{1}$$

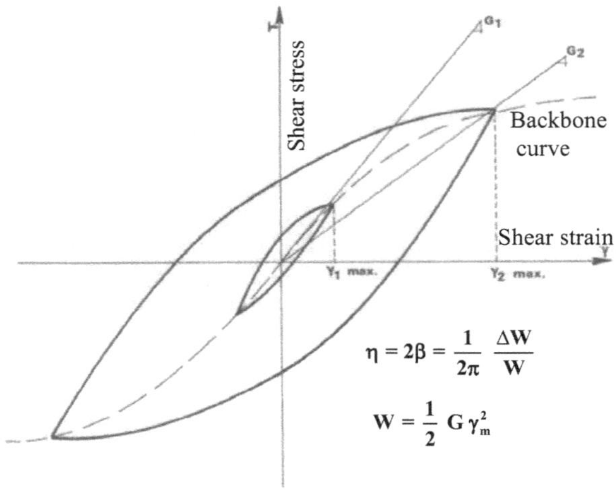

Figure 3. Shear stress-shear strain curves for constant amplitude cyclic loading

When cycles of constant amplitude are reproduced in the laboratory on a soil sample, the stress strain curves depicted in Figure 3 are obtained. In the (τ,γ) plane the behaviour is characterized by an hysteresis loop, the surface and inclination of which depend on the strain amplitude. The larger the shear strain the wider the hysteresis loop and the flatter it is on the horizontal axis. Furthermore, experimental evidence shows that the shape of the loop is not affected by the loading rate. As soon as the cycles have no longer a constant amplitude, the description of the behaviour becomes more complex. One example is depicted in Figure 4.

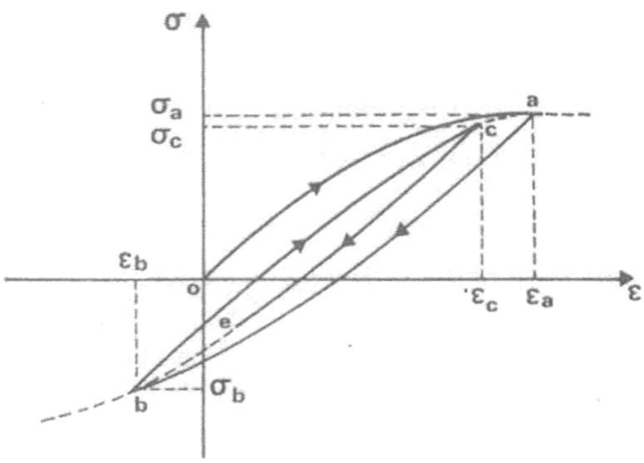

Figure 4. Arbitrary cyclic loading

Until point *b* is reached, the stress strain curve is identical to the one depicted in Figure 3 (first loading curve, also called backbone curve, followed by the unloading curve); at point *b*, such that $\sigma_b < \sigma_a$ the sign of the loading is reversed; the path is given by the curve *bc*, then eventually by *ce* if the loading sign is again reversed at *c*. If, on the contrary, the loading is continued beyond *c* the path is given by *ca* and then follows the backbone curve.

The shear stress strain behaviour described above is accompanied by volumetric strains (Figure 5).

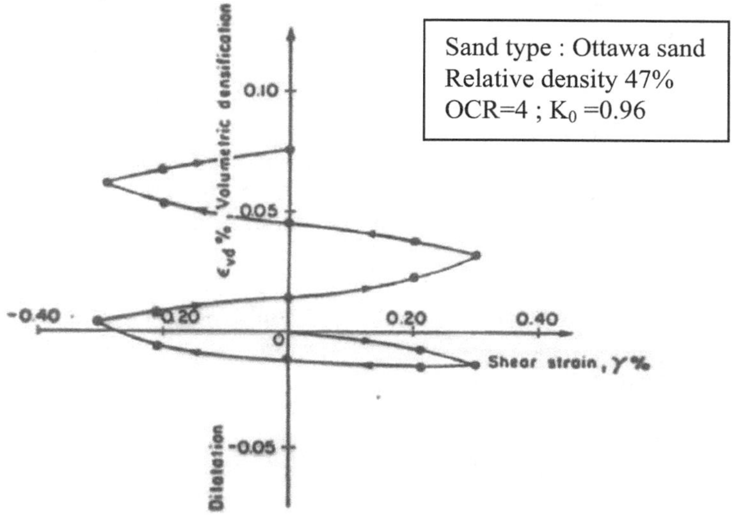

Figure 5. Volumetric strains under cyclic loading

These volumetric strains point out that the behaviour is no longer elastic, even non linear elastic. In a dry material they induce a hardening behaviour. Therefore, even for symmetric closed cycles with identical strain amplitudes, the hysteresis loop measured during, for example, the fourth cycle is different from the loop obtained during the first cycle. The latter is less inclined on the horizontal axis and exhibits a smaller area. For an impervious saturated material, strains occur under constant volume conditions because the pore water does no have time to drain from the sample. The volume change tendency is compensated by a pore pressure build up, hence by a decrease of the effective stresses.

The (over)simplified examples described above illustrate the complexity of soil behaviour, which is highly non linear and inelastic. The constitutive model adopted in practice, and described hereafter, only takes into account the deviatoric behaviour (Figure 3); the volumetric changes are often neglected, at least for soil structure interaction analyses, except in the case of true elasto-plastic constitutive laws.

3 Modelling of Soil Behaviour

A complete description of the behaviour is obtained if, starting from an equilibrium state characterized by a stress field $\underline{\underline{\sigma}}$ and an associated strain field $\underline{\underline{\varepsilon}}$ it is possible, for any strain increment $d\varepsilon$ (or any stress increment $d\underline{\underline{\sigma}}$), to determine the new stress field (respectively strain field) corresponding to a new equilibrium state. In the most general situation, time is also a variable that must be taken into consideration in the constitutive law; however, for most soils, this parameter can be neglected because soils are not rate dependent materials in the range of loading rates induced by earthquakes. The development of a complete constitutive relationship is the ultimate goal of soil modelling; however, in view of the complexity of the behaviour this task constitutes a real challenge and it can be stated that, at the present time, there does not exist a universal constitutive model. Every model available in the literature has its own advantages but also its drawbacks and limitations.

Facing that challenge the earthquake geotechnical engineer often favours, in engineering practice, a more straightforward approach, traditional in soil mechanics. According to this approach, the loading path to which the soil element will be subjected during an earthquake is anticipated and reproduced in the laboratory, or possibly in the field. The parameters measured during those tests are then directly used in the computations. For instance, in soil mechanics, the settlement of a finite thickness compressible clay layer under a wide spread load is studied from a one dimensional compressibility test with zero radial strain.

It must be realized that this approach is not similar to the development of a constitutive model, even if the measured stress-strain curves are represented by mathematical relationships. This kind of modelling remains valid only for the stress paths for which they were established, or for similar stress paths. Its extrapolation to fundamentally different stress paths is erroneous and not permissible. Furthermore, more than often, this approach is only an imperfect modelling of the actual physical phenomena; for example, the equivalent linear visocelastic model does not take into account the volumetric strains (settlements) that the soil experiences under shear loading. In addition, the stress paths duplicated in the experiments represent ideal, somewhat crude, representations of the actual paths. This kind of approach is a good compromise between the actual phenomenon to be modelled and its easy implementation. When used with care it can be a very powerful tool.

Before describing the experimental observations and their mathematical modelling, it is important to realize that, given the time scale of earthquake loading, most soils behave under undrained conditions during the earthquake. The soil permeability is not large enough (with respect to the rate of loading) to allow for drainage and dissipation of excess pore pressures. Consequently, the approach described previously is implemented in terms of total stresses; again, this implementation is an oversimplification of the actual situation as soil behaviour is governed by effective stresses.

Finally, in the rest of the chapter we will restrict ourselves to the description and modelling of the pre-failure behaviour of soils. The modelling of soil behaviour at failure is a matter of specific approaches. When a true constitutive law is available, that distinction is not required; the constitutive model allows for an accurate modelling of soil behaviour from very small strains (quasi elastic behaviour) to very large strains associated with failure.

For a more detailed description of soil behaviour, the reader can refer to Hardin (1978), Pecker (1984) or Prevost (1998).

As evidenced by the experimental observations described in paragraph 2, the soil cannot be modelled with a linear constitutive relationship, at least as soon as strains become significant. The strain thresholds for which non linearities appear are usually very small (10^{-6} to 10^{-4}). It is however fundamental to make a distinction between recoverable, or quasi elastic, strains and irrecoverable strains that develop for larger thresholds (10^{-4} to10^{-3}). The values of these two thresholds, which will be denoted γ_s and γ_v, depend on the nature of the material that can be roughly characterized by its plasticity index (Vucetic, 1984). Table 2 and Figure 6 delineate both domains as well as the mathematical description that can be used in numerical analyses.

Table 2. Strain thresholds for cyclic loading

CYCLIC SHEAR STRAIN AMPLITUDE γ		BEHAVIOUR	ELASTICITY and PLASTICIY	CYCLIC DEGRADATION in STAURATED SOILS	MODELLING
Very small	$0 \leq \gamma \leq \gamma_s$	Practically linear	Practically elastic	Non degradable	Linear
Small	$\gamma_s \leq \gamma \leq \gamma_v$	Non-linear	Moderately elasto-plastic	Practically non-degradable	Equivalent linear
Moderate to large	$\gamma_v \leq \gamma$	Non-linear	Elasto-plastic	Degradable	Non-linear

Strain amplitudes induced by major earthquakes in the European context are capable of creating significant non linearities, and possibly irrecoverable deformations ($\gamma \geq \gamma_s$ or γ_v). As indicated in Table 2 a different behaviour corresponds to each domain and must be characterized by specific constitutive parameters.

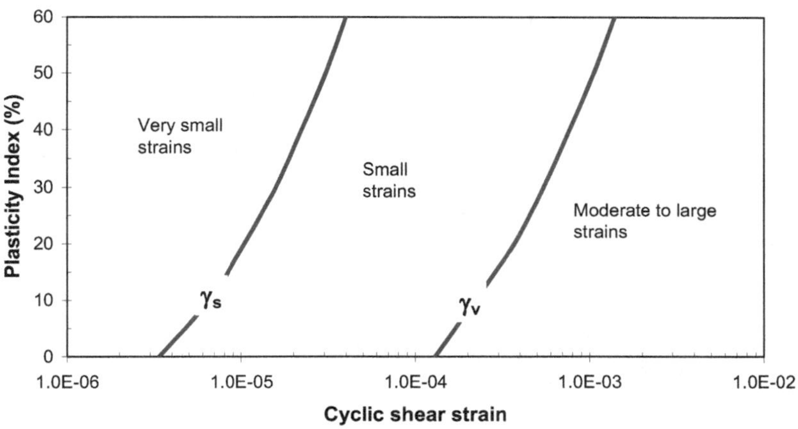

Figure 6. Threshold values for cyclic shear strains

3.1 Mathematical Description for Small Strains $\gamma \leq \gamma_s$: Elastic Model

Typically for strains smaller than $10^{-6} - 10^{-4}$ soils behave elastically. Some saturated materials may however exhibit some viscous energy dissipation. The natural soil constitutive model to use will therefore be the linear elastic, possibly viscoelastic, model. For an isotropic material the shear modulus G (equivalent to the Lamé coefficient μ of continuum mechanics) and the bulk modulus B completely describe the model. Alternatively, one can use the elastic wave propagation velocities V_S (shear wave) and V_P (dilatational wave), which are related to the previous quantities:

$$G = \rho V_S^2 \qquad B = \rho\left(V_P^2 - \frac{4}{3}V_S^2\right) \tag{2}$$

where ρ stands for the material mass density.

Characterization of soil behaviour within its elastic domain, although not very useful in earthquake engineering, is nevertheless important for a whole class of problems including machine vibration, geophysical measurements, etc…

3.2 Mathematical Model for Moderate Strains $\gamma_s \le \gamma \le \gamma_v$: Equivalent Viscoelastic Model

In that strain range, more or less pronounced non linearities and energy dissipation become apparent in the shear-strain curve. As linear viscoelastic model exhibits under harmonic loading hysteresis loops it is tempting to model the soil behaviour with such models (Figure 7). However, the viscoelastic model lends itself to an energy dissipation mechanism that is frequency dependent, in contradiction with experimental observation (paragraph 2).

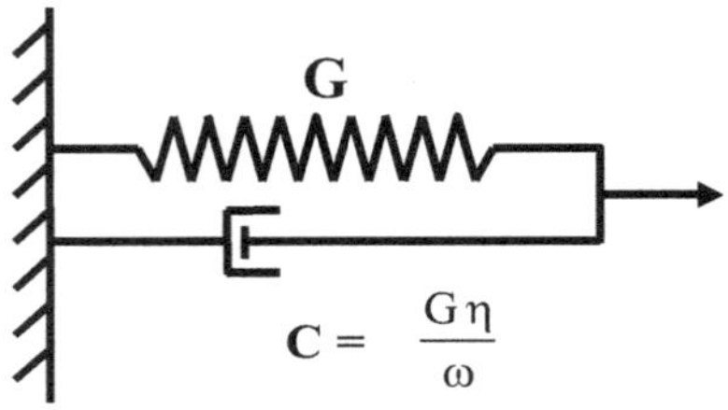

Figure 7. Viscoelastic model

To overcome that difficulty Jennings (1964) proposed to establish an equivalence between the actual material and the model, based on the equality of the energy dissipated in the material and at resonance in the model. For the Kelvin-Voigt model (Figure 7), the one dimensional shear stress-shear strain relationship writes:

$$\tau = G\gamma + C\dot{\gamma} \tag{3}$$

where G and C are the spring and dashpot coefficients; γ and $\dot{\gamma}$ are the shear strain and shear strain rate. For harmonic loading:

$$\gamma = \gamma_m e^{i\omega t} \tag{4}$$

equation (3) becomes:

$$\tau_m = G\left[1 + i\frac{C\omega}{G}\right]\gamma_m = G^*\gamma_m \tag{5}$$

where G^* is a complex valued modulus. The energy dissipated during one cycle of loading is equal to:

$$\Delta W = \pi C \omega \gamma_m^2 = \pi \operatorname{Im}\left(G^*\right)\gamma_m^2 \tag{6}$$

where Im(G^*) represents the imaginary part of G^*.

C being a constant it is obvious that the dissipated energy depends on the loading frequency $f = \omega / 2\pi$.

Referring to Figure 3, and without entering into details, it is possible to define for a material with rate independent energy dissipation a normalized relationship for the dissipated energy ΔW during one cycle of loading. The normalizing quantity is the elastic energy W stored during one cycle:

$$W = \frac{1}{2} G \gamma_{\mathrm{m}}^2 \tag{7}$$

the equivalent damping ratio β, or the loss coefficient η, are frequency independent

$$\eta = 2\beta = \frac{1}{2\pi} \frac{\Delta W}{W} \tag{8}$$

The dissipated energy during one cycle of loading, which is then frequency independent, can then be written:

$$\Delta W = \pi G \eta \gamma_{\mathrm{m}}^2 \tag{9}$$

Equivalent linear viscoelastic models are defined by a constitutive relationship (for one dimensional loading) of the type:

$$\tau_{\mathrm{m}} = G^* \gamma_{\mathrm{m}} \tag{10}$$

where G^* is a complex valued secant modulus that must be defined to yield the same stiffness and damping properties as the actual material. Several models have been proposed to achieve that purpose. Their characteristics are defined in Table 3.

Table 3. Characteristics of equivalent viscoelastic linear constitutive models

	COMPLEX MODULUS $G^* = \tau/\gamma$	DISSIPATED ENERGY IN ONE CYCLE ΔW	MODULUS $\lvert G^* \rvert$
MATERIAL		$\pi G \eta \gamma_{\mathrm{m}}^2$	G
MODEL 1	$G = [1 + i\eta]$	$\pi G \eta \gamma_{\mathrm{m}}^2$	$G\sqrt{1+\eta^2}$
MODEL 2	$G e^{i\theta}$ $\eta = 2\sin(\theta/2)$	$\pi G \eta \gamma_{\mathrm{m}}^2 \sqrt{1 - \frac{\eta^2}{4}}$	G
MODEL 3	$G\left[\sqrt{1-\eta^2} + i\eta\right]$	$\pi G \eta \gamma_{\mathrm{m}}^2$	G

The first two models were developed by Seed and his co-workers (Seed *et al*, 1970); the third one is due to Dormieux (1990). Examination of Table 3 shows that the first model adequately duplicate the dissipated energy but overestimates the stiffness; the second one duplicates the stiffness but underestimates the dissipated energy and the third one is the only one fulfilling both conditions. In standard practice, the most commonly used model is the second one.

An alternative representation of the data depicted in Figure 3 is obtained by plotting the secant shear modulus and the equivalent damping ratio (or the loss coefficient) as a function of the shear strain (Figure 8). This figure clearly shows the existence of the shear strain threshold beyond which the secant modulus is no longer constant, as well as its dependence on the plasticity index.

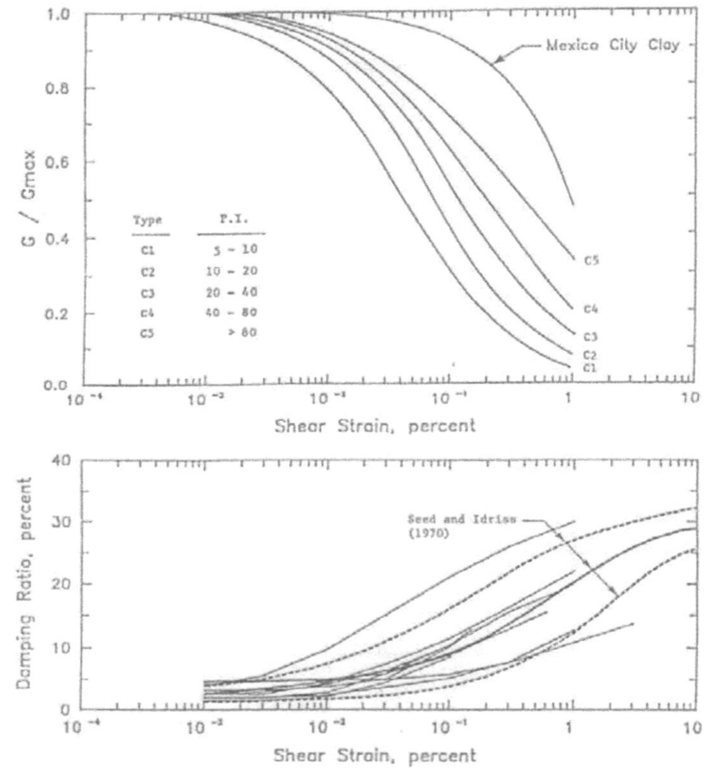

Figure 8. Typical variations with shear strain of shear modulus and equivalent damping ratios for clays (Vucetic & Dobry, 1991)

Correlatively, Figure 9 shows that for cyclic shear strains smaller than 10^{-4} to 10^{-3} the volumetric strain remains negligible, although shear non linearities have already occurred. As a consequence, in saturated impervious soils, significant variations in the pore pressure do not take place below this threshold. In that strain range, soil behavior is highly non linear, but approximately elastic; permanent changes in the soil microstructure remain negligible.

In the equivalent viscoelastic linear modeling described above, the soil characteristics (secant shear modulus and equivalent damping ratio) are chosen at variance with the "average" induced shear strain in order to reproduce, at least in an approximate manner, soil non linearities.

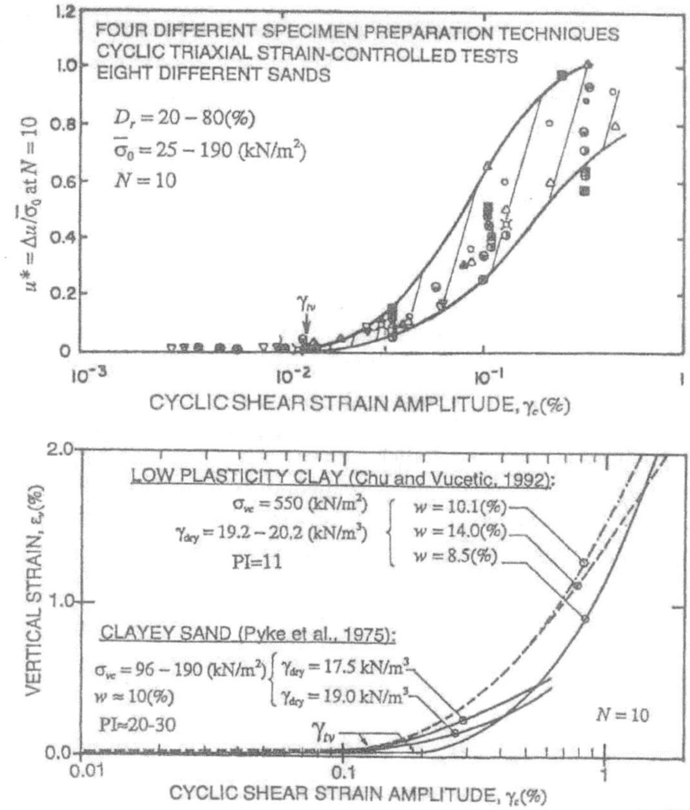

Figure 9. Pore pressure (top) and irrecoverable volumetric strain (bottom)

The results presented here above within the framework of one dimensional shear loading can be easily generalized to more general harmonic loading paths, bearing in mind all the limitations already mentioned on the validity of the constitutive model. The constitutive law is written as Hooke's generalized law:

$$\underline{\underline{\varepsilon}} = \underline{\underline{\underline{\underline{\Lambda}}}}^* : \underline{\underline{\sigma}} \tag{11}$$

where $\underline{\underline{\varepsilon}}, \underline{\underline{\sigma}}$ are the second order strain and stress tensors and $\underline{\underline{\underline{\underline{\Lambda}}}}^*$ the fourth order tensor of elasticity built from the two complex valued parameters (for an isotropic material) G^* and B^* (see Table 3) defined from the secant shear and bulk moduli and their associated equivalent damping ratios. In practice, the same damping ratio is used for both moduli.

The equivalent viscoelastic model described above is extensively used in engineering practice. When used in combination with an iterative process to choose properties (G, B and β) compatible with the average induced shear strain, they provide results for accelerations and stresses which are in good agreement either with observations or with more sophisticated (truly non linear) models. One of the greatest advantages of these models, beyond their simplicity, is the small number of required parameters: G, B, and β. Their main limitation is their limited range of validity and their inability to calculate irrecoverable shear or volumetric strains. To have access to these quantities more sophisticated models are required.

3.3 Mathematical Models for Large Strains $\gamma_v \leq \gamma$: Non Linear Models

In that strain range significant changes occur in the soil microstructure (grain rearrangement) inducing irrecoverable shear and volumetric strains (Figure 5 and Figure 9). These changes induce settlements in dry or unsaturated soils and pore pressure build up in saturated impervious soils. Pore pressures may rise to a condition where the effective stresses become equal to zero and, consequently, the soil shear resistance drops to a very small value; this phenomenon is known as liquefaction. Even if liquefaction is not reached, the pore pressure increase induces a drop in the soil stiffness (Figure 10). These two factors, shear stiffness degradation and loss of shear strength, make saturated soils subjected to strains larger than γ_v highly non linear. This behaviour, with irrecoverable strains, can only be described with non linear models. Past experience, experimental evidences and site observations reveal that elastoplastic models with hardening are the mot appropriate models (Prevost, 1978, 1987). These models must be written in terms of effective stresses; in saturated soils, coupling between the fluid phase and the solid phase needs to be taken into account within the framework of porous media mechanics (Prevost, 1980).

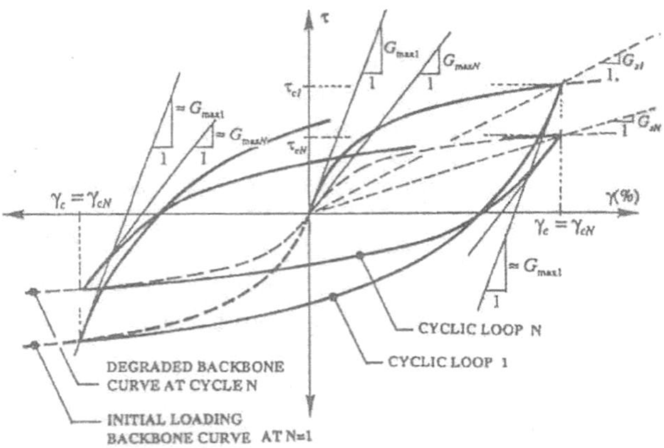

Figure 10. Stiffness degradation with the number of cycles

Except for exceptional situations, these non linear models are seldom used in engineering practice. The reasons stem from the difficulty to choose the most appropriate model (there is not a universal model), the difficulties in measuring the parameters entering the constitutive model, and the expertise required for running non linear dynamic analyses.

References

AFPS, (1990). Rapport de mission, Séisme de Loma Prieta.

Dormieux, L., Canou, J. (1990). Determination of Dynamic Characteristics of a Soil based on Cyclic Pressumeter Test , *3rd Int. Symposium on Pressumeters*, Oxford, 159-168.

Hardin, B.O. (1978). The nature of stress-strain behavior for soils. State of the Art, *International Conference on Earthquake Engineering and Soil Dynamics*.

Jarpe, S., Hutchings, L., Hauk, T., Shakai, A. (1989). Selected strong and weak motions data from the Loma Prieta earthquake sequence. *Seismological Research Letters*,60, 167-176.

Jennings, P.C. (1964). Periodic response of a general yielding structure. *Journal of the Engineering Mechanics Division,* ASCE, vol.90, n°EM2.

Pecker, A.(1984). *Dynamique des sols*, Presses de l'Ecole Nationale des Ponts et Chaussées.

Prevost, J.H.,(1978). Plasticity theory for soil stress-strain behavior, *Journal of Engineering Mechanics Division*, vol. 104, n° EM5.

Prevost, J.H. (1987). Modelling the behavior of geomaterials, *Course notes « Non linear soil mechanics and dynamic soil-structure interaction »*, Lausanne.

Prevost, J.H. (1980). Mechanics of continuous porous media. *Journal of Engineering Science,* vol. 18, 787-800.

Seed, H.B., Idriss, I.M. (1970). Soil moduli and damping factors for dynamic response analysis, *Report EERC 70-10. Earthquake Engineering Research Center.*

Seed, H.B., Ugas, C., Lysmer, J. (1976). Site dependent spectra for earthquake resistant design, *Bulletin of Seismological Society of America*, Vol. 66, 221-243.

Vucetic, M. (1984). Cyclic threshold strains in soils, *Journal of Geotechnical Engineering,* ASCE, Vol. 120, n°12, 2208-2228.

Vucetic, M., Dobry R. (1991). Effect of soil plasticity on cyclic response, *Journal of Geotechnical Engineering,* ASCE, Vol. 117, n°1, 89-107.

Determination of Soil Characteristics

Alain Pecker[1,2]

[1] Solid Mechanics Laboratory, CNRS UMR 7649, Department of Mechanics, Ecole Polytechnique, Palaiseau, France
[2] Géodynamique et Structure, Bagneux, France

Abstract. Determination of soil characteristics is a key aspect in constitutive modelling. The most sophisticated constitutive models are useless if the input parameters are incorrect or not properly defined. The strain range of interest for earthquake engineering goes from very small strains up to 10^{-3} and requires the use of especially dedicated tests. A combination of field and laboratory tests is the most effective way to achieve a reliable definition of oil behaviour.

1 Introduction

The determination of soil characteristics constitutes one of the key aspects of Soil Mechanics, in general, and more particularly of Soil Dynamics. The most sophisticated constitutive models are useless if the input parameters are incorrect or not properly defined. If the theoretical aspects of Soil Dynamics have very often taken over the experimental ones, during the last two decades significant improvements have been made in the latter aspects and new measuring techniques have emerged. The most significant improvements deal with the determination of elastic or pseudo-elastic characteristics, i.e. in a strain range extending from 10^{-6} to approximately 10^{-4}. This range is typical of the strains encountered in earthquake engineering and is well distinct from the strain range covered in Soil Mechanics where strains smaller than 10^{-2} are seldom encountered. Alike in Soil Mechanics, laboratory and in-situ techniques are developed in parallel. In any case, these two different approaches are not exclusive of each other; they are more than often complementary and should be systematically used to achieve a reliable definition of soil behaviour. Each of them has its advantages, but also its limits, which must be known to ascertain its domain of application and validity.

The state of practice for the description of soil behaviour usually makes a distinction between the pre-failure behaviour and the behaviour at failure (liquefaction, for instance). Not all techniques allow for testing the soil up to failure. The field tests and some laboratory tests are restricted to the measurement of characteristics in the small or medium strain range and therefore are well-suited for the determination of the soil parameters entering the definition of the equivalent visco-elastic constitutive model used in everyday practice.

Among all the testing techniques, it is possible to make a distinction between those which allow for a direct measurement of a soil parameter and those which define this parameter from experimental correlations. Laboratory tests and some field tests (wave velocities measurements) belong to the first category while the second category is restricted to field tests (SPT, cone penetration tests,...); the experimental correlations between a given soil parameter and the measured property (blow count, penetration resistance,...) are established from the observed behaviour of a soil deposit during an actual earthquake. These correlations are mostly dedicated to the assessment of the liquefaction resistance of soils and, as such, will not be dealt with in this lecture. Finally, all the testing methods used for assessing the global behaviour of a structure, in a broad sense, are not treated hereafter. Among these methods, one can list the measurement of ambient vibrations to determine the vibrational characteristics of a soil deposit, shaking table or centrifuge tests used for the soil-structure interaction studies.

2 Field Tests

2.1 Generalities

Nowadays, all field tests for the direct measurements of a soil parameter are based on the measurement of wave propagation velocities (Ballard – Mac Lean 1975; Woods 1994). The principle consists in generating in the soil an impulse wave of a known type (compression wave P, shear wave S, Rayleigh wave R) and in measuring the time t required for the wave to propagate between two receivers separated by a known distance d. The wave propagation velocity is:

$$V = d/t \tag{1}$$

where V represents either V_P, V_S or V_R.

In an elastic isotropic medium, the wave propagation velocity is related to the two parameters describing the elastic constitutive behaviour:

$$G = \rho V_S^2 \tag{2}$$

$$\lambda + 2G = \rho V_P^2 \tag{3}$$

where G (shear modulus) and λ represent the Lamé parameters and ρ the soil mass density. It is often more convenient to relate V_S and V_P to the shear modulus G and to some other parameters which may be Poisson's ratio ν, bulk modulus B or Young's modulus E. These parameters are related to V_S and V_P by:

$$\nu = \frac{V_P^2 - 2V_S^2}{2\left(V_P^2 - V_S^2\right)} \quad , \quad B = \rho\left(V_P^2 - \frac{4}{3}V_S^2\right) \quad , \quad E = \rho\frac{3V_P^2 - 4V_S^2}{V_P^2 - V_S^2}V_S^2 \tag{4}$$

Except for the ideally homogeneous isotropic elastic medium, which does not exist, the interpretation of measurements in terms of Rayleigh waves is more difficult; the spectral analysis of surface waves (SASW) described hereafter is based on the measurement of this wave propagation velocity. In a homogeneous medium, the knowledge of V_R allows for the computation of V_S once Poisson's ratio is known; the ratio V_R/V_S varies between 0.92 and 0.96 when ν increases from 0.25 to 0.5.

In all tests, the mechanical device for generating the impulse wave does not involve a large energy and consequently the induced strains in the soil remain small and the soil behaves elastically. Equations (2) to (4), derived from the theory of elasticity, are applicable and the computed moduli of deformation are associated with the elastic, or small strain values (strains smaller than 10^{-6}).

2.2 Measurement of Material Damping

Theoretically, the variation of the wave amplitude between the receivers gives the wave attenuation. This attenuation is composed of two terms: one is related to the inelastic attenuation of the medium in which the wave propagates, which is small since the soil behaves quasi elastically, and one is related to the geometric attenuation arising from the expansion of the wave front from the source. These two terms are referred to as material damping and geometric (or radiation) damping respectively. The radiation damping is a function of subsurface topography (profile, discontinuities,...), wave type and distance from the source to the receiver; it is predominant and could theoretically be evaluated. In practice, in view of our partial knowledge of the subsurface topography, the accuracy of such a computation is poor and does not allow for the evaluation of the material damping which is an order of magnitude smaller than the radiation damping.

To avoid the explicit computation of the radiation damping, Stewart and Campanella (1993) have proposed an interpretation method, named Spectral Ratio Slope (SRS). The material damping is computed from:

$$\beta = \frac{kV_S}{2\pi} \quad , \quad k = -\frac{\partial^2}{\partial f\, \partial z}\left[\mathrm{Ln}\left(\frac{A_R}{A_0}\right)\right] \tag{5}$$

in which the double differentiation (with respect to frequency f and depth z) eliminates the contribution of radiation damping. In equation (5), A_0 and A_R represent the Fourier transform amplitudes of the reference and recorded signals at depth z where material damping is computed.

2.3 Determination of Propagation Time

All methods require an accurate determination of the propagation time t of a wave between two receivers (equation(1)). Two techniques can be used:

- Direct measurement from the temporal signals;
- Indirect measurement from the cross-correlation function of the signals, computed either in the time domain or in the frequency domain.

Direct measurement

Traditionally, direct measurements are more often used. The main difficulty lies in an accurate identification of the time of arrival of a given wave. If for the compression wave P, the fastest one, the identification is easy, the situation is trickier for the shear wave S which reaches the receiver embodied in the wave train. Equations (2) and (3) point out the advantage of picking up the shear wave arrival, which allows for a direct computation of the shear modulus. In order to favour the formation of shear waves, mechanical sources have been developed which must be preferred to explosive sources; the small imparted energy developed by these mechanical sources does not constitute a limitation of the method since it is always possible to sump up many signals to produce signals with a high signal to noise ratio. Figure 1 illustrates the direct determination of a shear wave propagation time between two receivers located in adjacent boreholes.

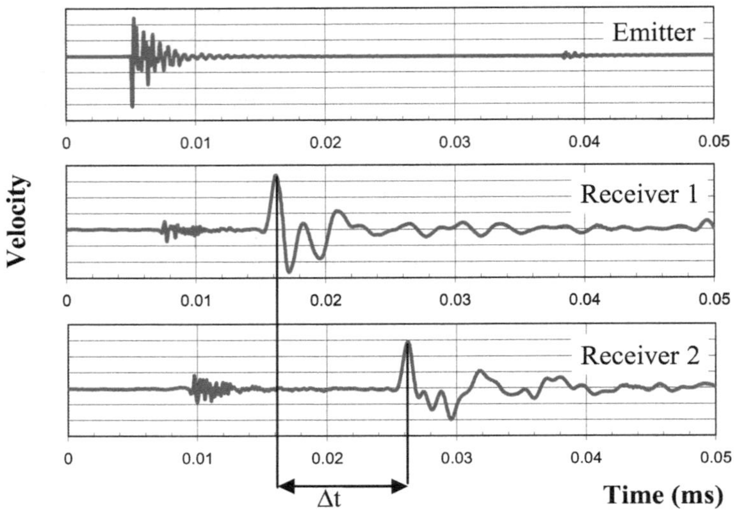

Figure 1: Direct determination of shear wave travel time

In order to facilitate the detection of shear waves, some mechanical sources offer the possibility of inverting the direction of the impulse. This inversion induces a different polarity for the shear wave, the detection of which on the traces of the signals is easier; the polarity of the compression wave is not affected.

Indirect measurement

With the development of computers, data processing techniques allow for a more accurate determination of the wave propagation time, which, in addition, can be automated. These indirect determinations are based on the computation of the cross-correlation function of two signals. Let $g(t)$ and $h(t)$ be two records obtained simultaneously at two receivers; their cross-correlation function is defined by:

$$\Re(\tau) = \int_0^T g(t)h(t+\tau)\,\mathrm{d}t \tag{6}$$

where T is the total duration of the signal and τ the time lag. If both functions $g(t)$ and $h(t)$ are proportional, but if one is shifted by a quantity t^* with respect to the other, the cross-correlation function $\Re(t)$ is maximum for $\tau = t^*$. This value t^* represents the time required for the wave to propagate from one receiver to another. In practice, the wave forms at both receivers are not identical because of dispersion and inelastic attenuation but the cross-correlation function of both signals exhibits a maximum at a time approximately equal to the wave propagation time. Figure 2 presents the cross-correlation function of both signals of Figure 1; the maximum is obtained for t=11.64msec, to be compared to 11.55msec read off the traces from a direct determination, i.e. a difference of 0.8%.

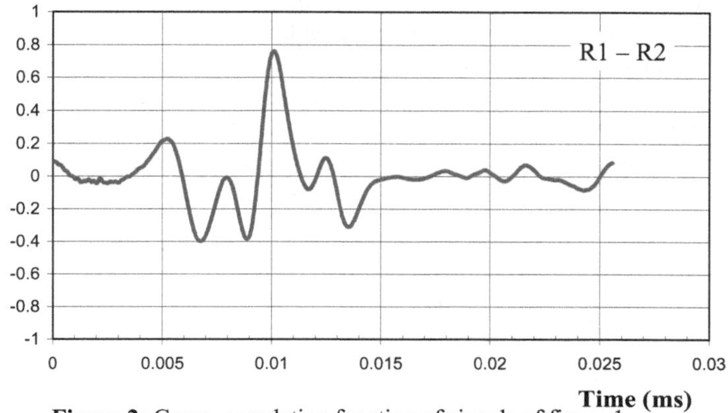

Figure 2: Cross- correlation function of signals of figure 1.

2.4 Distance Measurement

Equation (1) shows that the accuracy of the wave velocity determination V is related to the quality of the evaluation of t and of the distance between the receivers. This last measurement does not pose any particular difficulty when the receivers are at the ground surface; on the other hand, for in-hole measurements, it is mandatory to measure the verticality of holes to accurately know the exact distance between receivers at the depth of measurement; even high quality boreholes can deviate by 1% to 2%. It is worth noting that for some in-hole measurement techniques, suspension logging, the distance between receivers is fixed and therefore is not a cause of error.

3 Field Measurement Techniques

The geophysical tests can be classified in two categories:

- Tests realized from the ground surface, such as seismic refraction or spectral analysis of surface waves. These tests have the definite advantage of being easy to realize and have a comparatively low cost. The counterpart is the difficulty in the interpretation of the results which either requires crude schematization of the medium or very sophisticated numerical modelling.
- Tests realized in holes or between holes. These tests are more difficult to implement and have a higher cost, but they give more comprehensive and easier to interpret information.

During the last decade, in-situ testing techniques have moved very fast towards a reduction in cost and easier implementation; this trend is equally applicable to measurements from the ground surface (SASW) and to in-hole measurements (seismic cone). They should allow for the determination of the dynamic soil parameters on a routine basis, even for projects of secondary importance.

3.1 Tests Realized from Ground Surface

Seismic refraction

This method is well suited for preliminary site investigations but its implementation could be difficult. Theory and exploration techniques are well known and have been described in details in textbooks (Richart et al, 1970). The method consists in determining the propagation time of body waves (P or S) generated by a surficial source to a linearly aligned network of receivers. According to the soil profile beneath the receivers, the waves propagate either directly towards the receivers, or along refracted paths at interfaces between layers. For a three layers soil profile (two soil layers overlying a bedrock), the refraction paths are indicated in Figure 3.

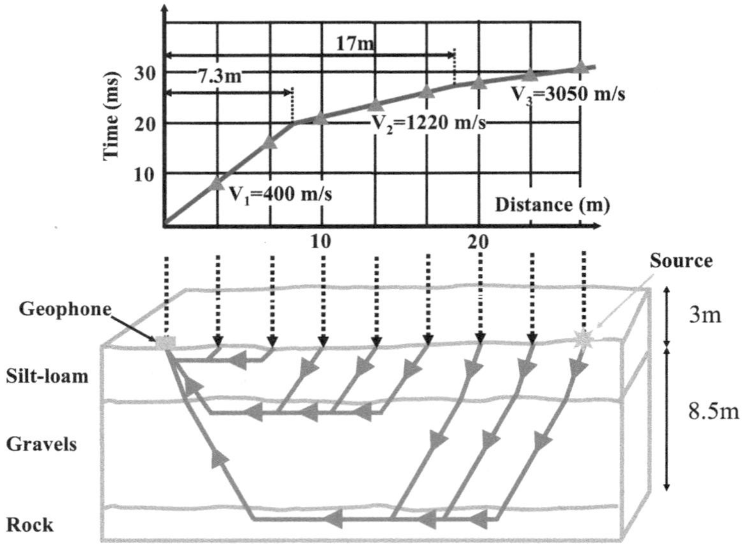

Figure 3: Seismic refraction

The slopes of the straight lines obtained by plotting the propagation time versus the distance are equal to the wave velocities. The source should preferably give rise to shear waves, which is achieved by horizontally hitting a block laid on the ground and vertically loaded. With this technique, an almost pure shear wave propagates perpendicularly to the source. Interpretation of the tests is done assuming that the wave velocity increases with depth. If this condition is not satisfied significant errors can affect the interpretation. Likewise, the presence of a soft layer (with a smaller wave velocity) entrapped between two stiff layers cannot be detected. Finally, in this method, the waves propagate near the layers interfaces; it is therefore impossible to sample an average wave velocity for the layer under consideration.

Spectral analysis of surface waves

This method represents an extension, and an improvement, of the harmonic vibration of a block foundation resting on the ground surface (Stokoe – Nazarian, 1985). In this older method, the wave length λ_R of a Rayleigh wave generated by the block foundation vibrating at frequency f is measured with a geophone moved on the ground surface. In a homogeneous soil profile, this velocity is frequency independent. In a layered profile, the wave propagation is frequency dependent and the relationship between frequency and wave velocity is called the dispersion curve; this phenomenon arises from waves with different wave lengths propagating in different layers: high frequency waves (short wave lengths) propagate near the layers interfaces; low frequency waves propagate in shallow and deep layers. Altering the vibration frequency allows to sample different depths and to derive a Rayleigh wave velocity profile and, if Poisson's ratio is known or estimated, a shear wave velocity profile.

The SASW method takes advantage of a more refined analysis of the records for the layered profile and of an easier and faster implementation. The generation of Rayleigh waves is produced by an impulse loading at the ground surface. Recorded signals are digitized and transferred into the frequency domain, the phase shift $\phi(f)$ between two recorded signals is determined, from which the propagation time and the wave propagation velocity are computed. Repeating these steps for different frequencies, the dispersion curve can be established. Figure 4 presents a dispersion curve constructed from the SASW method (Stokoe et al, 1994).

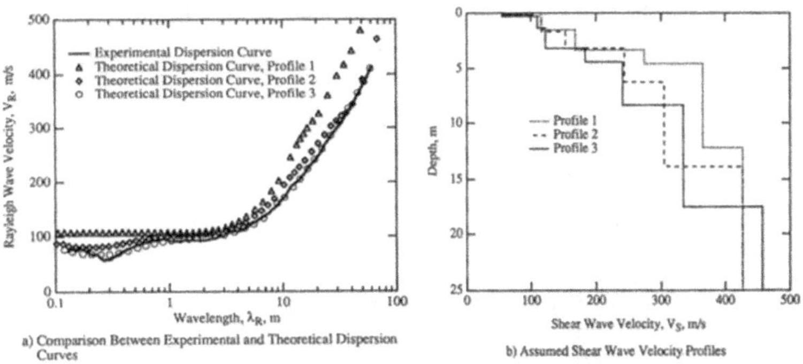

Figure 4. SASW test

The theoretical curve is compared to the experimental one and the wave velocity profile altered until a good agreement is reached between both curves. The wave velocity profile computed at the last iteration is deemed to represent the actual velocity profile. This procedure is illustrated in Figure 4 (Stockoe et al, 1994) which shows the evolution of the velocity profile with the iterations and the measured dispersion curve. The final wave velocity profile has been compared to measurements from a nearby cross-hole test, validating the results. However, the SASW method yields a stepped wave velocity profile as opposed to the more continuous variation obtained from a cross-hole test; this arises from the sampling method which is more global in the SASW method.

3.2 In-Hole Testing

In-hole tests can be grouped into two categories: those which require the prior drilling of borehole(s) in which measuring devices are lowered down (cross-hole, down-hole, suspension logging) and those for which the measuring device is lowered with the drilling tool (seismic cone). For the first category, tubing of the hole with a PVC casing is required and the annular space between the casing and the hole must be grouted to ensure a good coupling with the surrounding soil. These operations are time and cost-consuming; in addition, as previously mentioned, it is absolutely mandatory to measure the holes inclinations.

Down-hole tests

The measurement of the wave propagation velocity is performed along one hole. The source is placed at the ground surface and the receivers are placed inside the casing. The use of adequate sources (horizontal impact of a block foundation) gives rise to predominant shear waves. In its simplest version, only one receiver is used and lowered in the hole at successive depths. More elaborate techniques take advantage of a set of receivers clamped on the same frame which allow for a simultaneous recording of the same input signal. The accuracy in the determination of the wave velocity is better because the distances between receivers are perfectly known and because the signals are issued from the same impulse. Velocities measured in a down-hole test correspond to values along the vertical wall of the hole, for a vertically propagating wave. In-plane variations of these velocities cannot be assessed. On the other hand, with a sufficiently small spacing between the receivers, it is possible to sample layers with lower characteristics, even if they are entrapped in between two stiffer layers. This is one of the major advantages of the method. Figure 5 presents an example of a down-hole record.

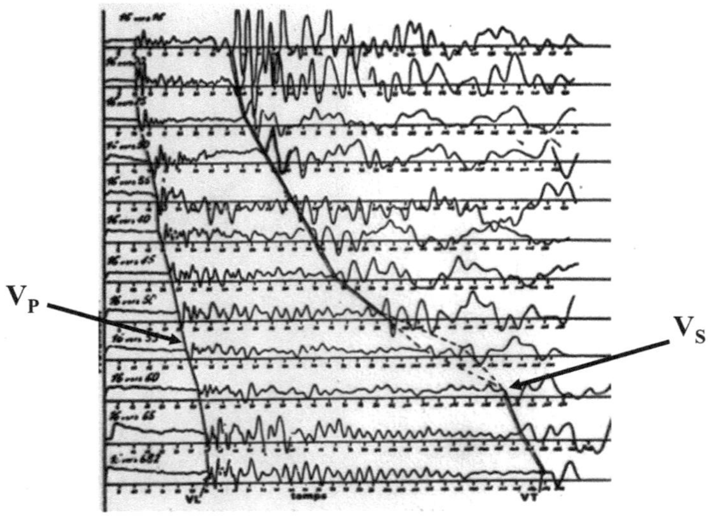

Figure 5. Down-hole record

Suspension logging

This method, for the measurements of P and S wave velocities, uses only one, preferably uncased, hole. The borehole is filled with water or drilling mud. The receivers and the source are clamped on the same frame and lowered down at the same time (Figure 6). The source, with the capability of inverting the wave polarity, is made of a horizontally oriented solenoid which generates a pressure wave in the fluid. When hitting the borehole wall, this pressure wave induces a radial displacement in the soil which gives rise to radially and vertically propagating body waves in the soil. The propagation and displacement of the boreholes wall in turn induce pressure waves in the fluid which are recorded by biaxial geophones. The separating distance between two geophones is 1m which permits an accurate resolution of the signal. The frame with a total length of 7m is suspended with a wire line from the ground surface; this test set-up makes possible the investigation at great depths, in excess of 100m (Nigbor – Imai, 1994).

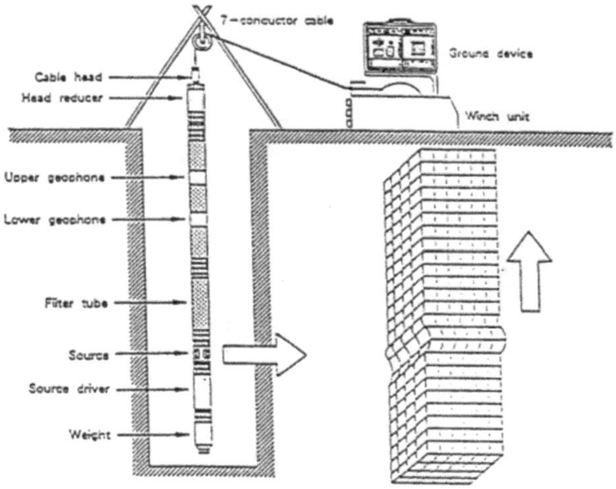

Figure 6. Suspension logging test set-up

Seismic cone

This is the only in-hole measuring technique which does not require the prior drilling of a borehole. The equipment is similar to the one used for cone penetration tests (CPT); in addition to the gages for measuring the point resistance, the skin friction, possibly the excess pore pressure, the cone is provided with a piezo-resistive gage for the detection of seismic waves and with inclinometers for the measurement of verticality (Campanella and Stewart, 1992). The seismic wave is generated at the ground surface by any appropriate mechanical device that favours the generation of shear waves. Recordings of the signals are made at stops during penetration, every 0.5m or 1m. In order to compare the amplitudes of the signals at various depths and to compute the material damping (§ 2.2), the source must be capable of generating reproducible signals. The test interpretation is straightforward and resorts to the computation of the propagation time of the wave between two successive positions of the receiver.

The major advantage of the test is, besides its low cost and quick realization, to allow within the same test for the determination of different parameters that can be used to characterize the site from a geotechnical standpoint: point resistance, friction ratio, pore pressure ratio, permeability, and wave propagation velocities. The limitations of the test are similar to those of the down-hole test:

good definition of the velocity profile only along the hole, limitation of the sampled depth because of the limiting capacity of the source.

Cross-hole test

Unlike the other in-hole techniques, the cross-hole tests require the drilling of at least two cased boreholes equipped with PVC tubings, grouted to the soil. To improve the test quality, it is better to use three, or more, aligned boreholes (Figure 7), spaced by a few meters (typically 3m to 10m). The impulse is imparted in the emitting borehole and the generated waves are recorded in the receiving boreholes with triaxial geophones located at the same depth as the source. The source and the receivers are lowered in the holes and clamped to the casing with pneumatic or hydraulic packers.

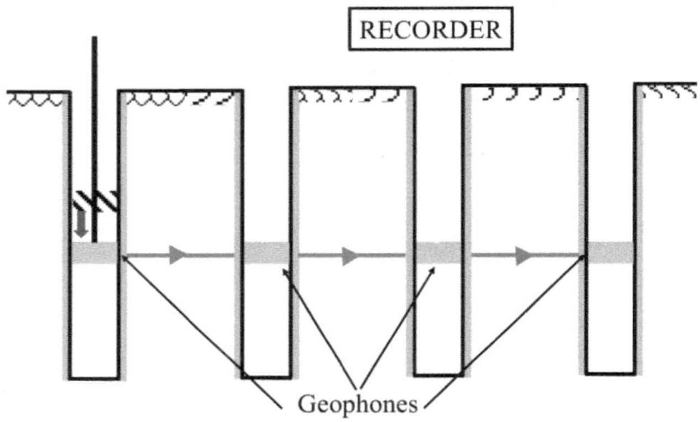

Figure 7: Cross-hole test set-up

The most commonly used sources are mechanical hammers with a vertically oriented stroke imparting a shear displacement to the borehole wall; this stroke gives primary rise to shear waves propagating horizontally in the soil. This type of sources allows for an inversion of the direction of the stroke, which modifies the polarity of the shear waves; in addition, it is possible to sum up the signals from different, reproducible, strokes. Use of at least two receiving holes present many advantages:

- the propagation time is computed from differential measurements between receiving holes, while with a two holes set-up, it is computed between the source and the receiver; the coupling between the emitting source and the casing being weaker than the coupling between the geophones and the casing, the determination of the initial time (time zero) may be subjected to errors. When only two holes are used, it is mandatory to have a geophone stick to the source;
- identification of refraction paths in stiffer layers located at shallow depths below the measuring depth is possible. Refraction paths in stiffer layers are a common source of error in cross-hole tests (Stokoe and Hoar, 1978).

The major advantage of cross-hole tests with respect to other in-holes measuring techniques is the possibility of obtaining mechanical characteristics within a horizontal plane and not along a borehole. Owing to the geological deposition of sedimentary deposits, the variation of soil characteristics is more pronounced in the vertical direction than in the horizontal one and the cross-hole test is therefore well-suited. If the receiving boreholes are arranged along different azimuthal directions (usually two mutually perpendicular directions), it is possible to measure the in-plane anisotropy of a layer; this possibility might be interesting in highly tectonized rock formations.

The cross-hole test, owing to its high versatility, and despite its relatively high cost, is the most powerful used test for important structures. Depths as deep as 100m can easily be reached in cross-hole tests.

3.3 Conclusions on field tests

All the previously described field techniques require, for an accurate interpretation, an accurate knowledge of the soil profile: for the cross-hole tests, it prevents from mistakes in the interpretation; for SASW it gives an initial guess for the iterative calculations; for all the tests, it helps defining a measuring depth increment adapted to the soil profile. It is therefore of paramount importance, for an accurate interpretation of the tests, to have a continuous log of a borehole close to the test. For the down-hole or suspension logging tests, it can be the testing hole; for the cross-hole test, it can be one of the holes.

In-situ testing techniques differ by the possibilities they offer and the information they provide. On the one hand, depths reached with geophysical tests performed from the ground surface (seismic refraction, SASW, down-hole, seismic cone) are limited, typically of the order of 50m; on the other hand, techniques in which the source is lowered in the hole, maintaining a limited distance between the source and the receiver (suspension logging, cross-hole), may sample larger depths, in excess of 100m.

Techniques involving only one hole (suspension logging, down-hole, seismic cone) give only local information on the soil characteristics, those of the soil in the immediate vicinity of the hole; however, they allow for an accurate definition of the profile. Seismic refraction or SASW tests provide global information on the medium, but do not allow for an accurate definition of the soil profile. Cross-hole tests are a good compromise giving average in-plane characteristics, at a scale of approximately 10m, while maintaining a good accuracy for the definition of the soil profile.

The main limitation of field tests, besides their inability, in the present state of practice, to induce large strains in the soils and therefore to test the soil behaviour beyond its elastic range, lies in the fact that the characteristics are evaluated under the stress field prevailing in the soil profile at the time of testing. More than often, constructions involve significant modifications to the stress field in the soil: these variations are related to excavations, backfills, modification of the hydraulic regime, stress increase due to buildings… Since elastic soil properties (shear modulus, for instance) depend upon the effective stresses supported by the soil, these modifications to the initial stress field lead to significant variations of the soil mechanical properties which must be assessed. It is therefore necessary to evaluate the in-situ stress field at the time of realization of the tests and complement the tests with laboratory tests in which variable stress fields can be imposed to the sample. For projects of secondary importance, with a limited budget, the variation with stress of the soil properties can be assessed on the basis of the classification tests, together with experimental published correlations on similar materials.

Finally, it is mandatory to complement the in-situ geophysical tests with an accurate description of the materials and with laboratory tests including, as a minimum, classification tests (water content, dry unit weight, grain size distribution, Atterberg limits,…) and, better, tests characterizing the general soil behaviour (compressibility tests, triaxial tests) and the cyclic and dynamic behaviour. In addition, the geophysical investigation should give an estimate of the horizontal and vertical effective stresses in the soil at the time of the survey to allow for a correct interpretation of the tests and for an extrapolation of the results to other stress conditions.

4 Laboratory testing

Some aspects of dynamic soil behaviour are easier to study in a laboratory, under controlled stress conditions. In addition, a comprehensive understanding of the soil behaviour under cyclic loading requires the realization of numerous tests carried out under various stress conditions and load paths. These conditions can only be achieved in a laboratory. However, to be representative of the actual soil behaviour, these tests have to be performed on truly undisturbed samples, which are capable of restoring the past history of the deposit in terms of strain and stress paths. This stress-strain history induces, from the microstructure standpoint, specific grains arrangements and bonding conditions between the grains. Since it is not conceivable to reproduce this microstructural arrangement, one has to rely on truly undisturbed samples which preserve this structure. This requirement of using truly undisturbed samples is the main difficulty which has to be faced for laboratory testing.

If sampling techniques of fine, cohesive materials (clay, mud, clayey silts) are well-established and efficient, it has always been a challenge to sample cohesionless, uniformly graded, materials. During the last decade, new techniques have emerged but their costs remain high for common projects. Among these techniques, one can list the freezing technique which has given satisfactory results and a new one, still in an experimental phase, which consists in filling the voids, by a temperature rise, with natural polymer, Agarose (Sutterer et al, 1996). Both techniques appear capable of preserving the particles arrangement during sampling and setting in the laboratory apparatus; soil disturbance is reduced to a minimum.

Among the laboratory tests, it is possible to make a distinction between the tests which can apply only one stress path from those which can impose various stress paths. As a rule, the first category of tests have been developed to duplicate as closely as possible the stress-strain path followed in-situ by a representative soil element. In those tests, the testing procedure is almost always the same: the undisturbed sample is consolidated under a (isotropic or anisotropic) stress state. For the test, this state of stresses is not necessarily identical to the in-situ stress state; however, for the interpretation of the test and its transposition to field conditions, it is necessary to evaluate this state of stresses with respect to peculiar states (in-situ stresses, preconsolidation pressure,…). After consolidation, the drainage valves are closed and the cyclic load is applied under undrained conditions. If the sample is saturated, no volume change takes place during loading; for unsaturated sample, some volume change takes place but its measurement is almost impossible. The testing procedure (consolidation followed by undrained loading) is close to actual field loading conditions. As a matter of fact, in the frequency range of interest (0.5Hz to 10Hz), most soils can be considered as loaded under undrained conditions. Resonant column tests and cyclic simple shear tests belong to this category.

The second category of tests have the capability of testing samples under various initial stresses and stress (or strain) paths, still preserving a uniform stress conditions within the sample, a good control on the volume change,… These tests can be used for the determination of parameters entering a general stress-strain constitutive law. On that account, the environmental testing conditions may widely differ according to the objective: tests may be performed under drained or undrained conditions with pore pressure measurement. Cyclic torsional shear tests on hollow cylinder and cyclic triaxial tests belong to that category.

Generally speaking, it is also possible to group the laboratory tests into three categories (Wood, 1978): free vibration tests, resonant tests and forced vibration tests for the measurement of hysteresis loops. The latter tests are usually performed at much lower frequencies (or the order of 1Hz) than the resonant tests; they are the only ones capable of loading the sample up to failure. Measurements of propagation wave velocities are seldom used for soil samples. This classification is adopted in the following.

4.1 Free Vibration Tests

The basic principle of the test consists in applying an initial displacement to the sample and to let it come back to its initial position in free vibration. According to the initial deformation, the vibration can be longitudinal, transverse or torsional. Measurement of the vibration frequency and of the attenuation of the vibration amplitude, allows for the computations of a modulus of deformation and of the logarithmic decrement which is related to the material damping. With this method, soil characteristics can be measured in a strain range extending from 10^{-5} to 10^{-2} (Seed – Idriss, 1970). In everyday practice, this technique is seldom used.

4.2 Resonant Tests

The basic principle of the test consists in applying forced vibrations to the sample and to tune the frequency of excitation until resonance occurs. The applied vibration can be longitudinal, transverse or torsional. Transverse vibrations can be applied with shaking tables and longitudinal or torsional vibrations with resonant column apparatus. A simplified sketch of a resonant column apparatus, which has been studied in details by Drnevich (1977), is given Figure 8.

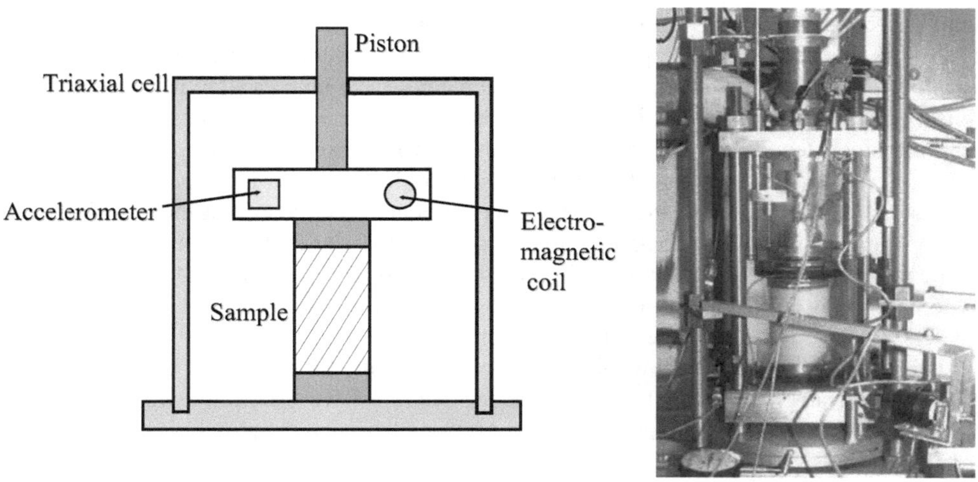

Figure 8: Resonant column test

The load is applied with electrical coils placed in a magnetic field. The input frequency of the alternating current is modified until resonance (for the first mode) occurs. The sudden stop of the applied load allows the sample to come back to its initial position under free vibrations; the material damping can be computed from the attenuation of amplitudes. Alternatively, material damping can be computed from the phase shift between the applied force and the induced displacement. Knowledge of the resonant frequency and of the associated mode shape (generally the fundamental one) is sufficient to compute the modulus of deformation, provided the constitutive relationship is fixed, a priori. In the resonant column test, in view of the small induced strains, elastic, or quasi-elastic, soil behaviour is assumed. For torsional loading, if θ denotes the rotation angle of the sample cross-section, J_0 and J the mass moment of inertia of the top mass and of the sample, ℓ the sample height and V_S its shear wave velocity, the frequency equation of motion is:

$$\frac{J}{J_0} = \frac{\omega \ell}{V_S} \tan \frac{\omega \ell}{V_S} \tag{7}$$

In the testing procedure, the input frequency is increased from a small value, of the order of 10Hz, until resonance is detected, either with a velocity transducer or an accelerometer placed on top of the sample. In the first case, the figure produced on an x-y oscilloscope is a straight line; in the second case, it is an ellipse with horizontal and vertical axes. At resonance, the displacement at the top is obtained from the accelerometer (or velocity transducer) and the shear strain computed. Increasing the current intensity in the electromagnetic coils increases the induced strain. Material damping β is computed either in steady state vibration from the phase shift ϕ between the force and the displacement or from the logarithmic decrement δ measured under free vibration after switching off the power (eq.(8)).

$$\tan\phi = 2\beta \quad , \quad \delta = 2\pi\beta \tag{8}$$

The resonant column test is as flexible as the cyclic triaxial test with possibilities of drainage control, pore pressure measurement, large range of consolidation stresses. On the other hand, the load path is unique: either a torque to measure the shear modulus G, or a longitudinal force to measure the Young's modulus E. Commercial apparatus have both capabilities, with successive applications on the same sample of both types of loading; they can therefore be used to evaluate all the elastic parameters, at least for an isotropic material. To compute the elastic parameters, only the resonant frequency and the geometric characteristics of the sample of the apparatus are required; neither strain nor displacement, measurements are necessary. The accuracy is therefore higher than in a test where the modulus is computed from force and displacement measurements.

The resonant column test measures the soil characteristics in a strain range included between 10^{-6} to approximately 5 10^{-4} for torsional tests and at smaller strains in axial tests. With hollow cylinders, strains up to 10^{-2} can be reached (Anderson, 1974). These latter tests have the advantage of creating a uniform strain field within the sample, but render the sample set-up extremely difficult. For strain amplitudes smaller than 10^{-4}, most soils behave elastically and the test is non destructive; it is then possible to perform several tests on the same sample by varying the ambient conditions (stresses, temperature,…).

To conclude, it is worth noting that resonant column tests are accurate and reproducible. Skoglund et al (1976) have shown that results obtained from six different laboratories, with different apparatus, but with the same soil material and testing procedures, are in good agreement. At the present stage, the test has been standardized (Drnevich et al, 1978).

4.3 Forced Vibration Tests

These tests were originally intended to duplicate as closely as possible, in the laboratory, the stress (or strain) paths followed by a soil element in the field; the cyclic simple shear test has been developed for that purpose. In these tests, a known cyclic stress (or strain) is applied to the soil sample and the induced strain (or stress) is measured. Typically, tests are performed at frequencies of the order of 1Hz (0.3 to 2Hz). The hysteresis loop is determined and, according to the stress path, the relevant parameters for the characterization of the soil behaviour are computed: for instance, in a simple shear test, the secant shear modulus and the equivalent damping ratio for the visco-elastic linear equivalent soil constitutive model. Nowadays, with the increasing development of more and more sophisticated constitutive relationships, numerous and complex parameters are required. Measurement of these parameters has become possible with the significant technological advances in the apparatus, test driving and servo-control equipment, data acquisition systems provided by micro computers.

Cyclic triaxial tests or torsional cyclic tests on hollow cylinders have the capability of applying widely different complex stress paths, which makes possible the determination of other parameters than a modulus of deformation or a hysteresis loop. More generally, forced vibration tests are well-suited for measuring characteristics in a strain range extending to 5 10^{-4} up to failure. It is no possible with commercial apparatus to reach smaller strains, associated with displacements of a few microns for standard size laboratory samples (diameter of the order to 70mm). In fact, evaluation of a modulus requires the measurement of a force and of a displacement; this latter quantity, with usual LVDT's, is not accurate because contact deformations at the cap and base of the sample may reach a few microns. Only specially designed devices give access to smaller deformations. As a general rule, forced vibration tests are a good complement to resonant column tests for the evaluation of soil characteristics in a higher strain range. Figure 9 (adapted from Woods, 1978) summarizes the domain of validity of each test.

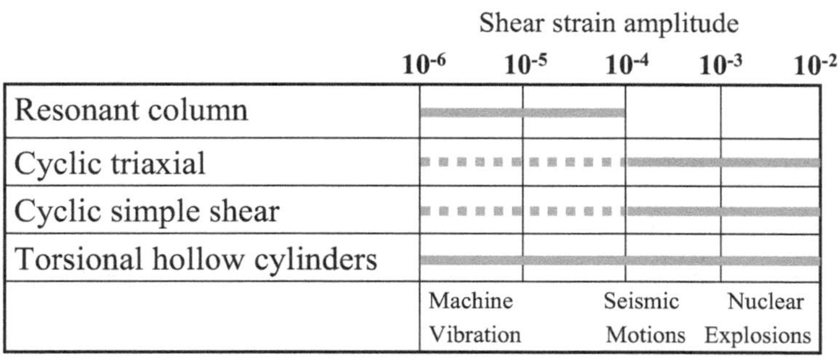

Figure 9 : Domains of application of laboratory tests

Cyclic triaxial test

The cyclic triaxial test has first been used by Seed and Lee (1966) and is presently the most widely used laboratory apparatus, especially for the evaluation of the cyclic strength. Conceptually, the apparatus is similar to the triaxial apparatus used in Soil Mechanics for monotonic tests; some adaptations are however required to increase the accuracy of the measurements, especially at small strains. One can mention: a stiffer frame, the absolute requirement of having the load gage located inside the triaxial cell to get rid of piston friction,... With these modifications, the cyclic triaxial apparatus presents all the needed versatility and most of the advantages one can expect from a laboratory equipment: good definition of an otherwise homogeneous stress field, possibility of saturation of samples, capability of applying isotropic or anisotropic consolidation stresses, pore pressure measurement devices,...To improve the accuracy of the strain measurements, El Hosri introduced proximity transducers located in the central third of the sample; the displacement measurement is done without any contact with an induction coil and a metallic target placed in the magnetic field of the coil. The accuracy on displacements is of the order of 0.2μm (strain of the order of 10^{-6}). Increasing the sample diameter from 30mm to 300mm, preserving the aspect ratio of 2, also increases significantly the accuracy, since the same strain, 10^{-5}, corresponds to à 0.8μm displacement for a 38mm diameter sample to 1.4μm for 70mm and to 6μm for 300mm (Dupas et al, 1988).

In everyday practice, the cyclic triaxial test is used for the evaluation of Young's modulus which can be converted into shear modulus provided Poisson's ratio is known, or for the undrained cyclic strength. In those tests, the sample is isotropically consolidated and subjected, under constant volume condition (undrained loading on a saturated sample) to an increase of the axial stress by a quantity $\sigma_d/2$. Cyclic triaxial tests can be performed under strain, force or stress controlled conditions. Young's modulus is calculated from the ratio of the axial stress to the axial strain ε. The shear modulus (secant modulus) and the shear strain γ are computed from these parameters. The equivalent damping ratio β is obtained either directly from the area of the hysteresis loop, or from the phase shift between the force and the displacement (eq. (8)). Data processing techniques described for field tests (cross correlation function) can be used to evaluate the phase shift.

The cyclic triaxial test is also used for the evaluation of the cyclic undrained strength of sands. In that case, the test is performed under force, or better stress, controlled conditions. The test is pursued until failure of the sample by liquefaction occurs; during the test, the applied stress, induced strain and pore pressure are continuously recorded. Cyclic triaxial tests are reliable and reproducible; cyclic undrained strengths measured on the same material with similar testing procedures in eight different laboratories using different equipment proved to be similar (Silver et al, 1976).

Cyclic simple shear test

The cyclic simple shear-test has been developed to study the stress-strain behaviour of soils under pure shear stress fields. Samples can be tested under plane strain conditions with possible rotation of the principal stresses during the test. It must not be confused with the direct shear test, developed by Casagrande, which is not suited for the study of the stress-strain behaviour. This test has long been considered as the test which most closely duplicates the field stress conditions in a soil element subjected to the vertical propagation of shear waves. The first tests under cyclic loads have been reported by Peacock and Seed (1968) and Silver and Seed (1971).

Presently, the most commonly used apparatus are derived from the original Norwegian Geotechnical Institute (NGI) device; the cylindrical sample is wrapped in a reinforced rubber membrane. The sample is consolidated under a vertical effective stress σ'_v and is radially confined by the reinforced membrane under a horizontal stress $K_0\ \sigma'_v$ (K_0, coefficient of earth pressure at rest). The cyclic shear stress is applied at the top horizontal plane of the sample with a hydraulic or pneumatic device; the membrane stiffness enforces a near simple shear deformation of the sample. Undrained tests are performed by keeping the volume constant, assuming that the vertical stress variation required to maintain the sample height constant is equal to the pore pressure variation that would be measured in truly undrained tests (De Groot *et al*, 1991).

Numerous numerical studies (Hvorslev – Kaufman, 1952; Roscoe, 1953; Duncan – Dunlop, 1969; Prevost – Hoëg, 1976) have pointed out the role played by the missing complementary shear stress on the vertical sides of the sample; to satisfy the equilibrium conditions, the resultant of the normal stresses on the top and bottom faces induces a moment which counterbalances the moment developed by the shear stresses on the same planes. Therefore, the distribution of shear and normal stresses is no longer uniform. However, experimental studies (Finn et al, 1971, Vucetic – Lacasse, 1982; De Groot *et al*, 1994) have shown that, except for the behaviour at large strains, beyond the peak shear stress, tests results do not seem to be affected by this lack of homogeneity. The use of the simple shear device is therefore justified. However, the versatility of the test is less than for the triaxial test (difficulty to control the volume change, no control on the radial stress, only one possible stress path,…). This test is however used either for the measurement of stress strain properties (shear modulus) or for the evaluation of the cyclic undrained shear strength.

Torsional cyclic shear tests

To ensure more homogeneous stress fields within the sample and to have control on the radial stress, Hardin et Drnevich (1972) have developed a torsional cyclic shear device to test hollow cylinders. Obviously, this apparatus cannot be used for testing undisturbed cohesionless samples and poses many difficulties for the set-up of cohesive samples. It is therefore not used in everyday practice. With the capability of executing on the same sample resonant column tests and forced vibration tests, this device gives access to the cyclic stress strain behaviour over the whole strain range of interest. It can also be used for the evaluation of the undrained cyclic shear strength.

References

Anderson D.G. (1974.). Dynamic modulus of cohesive soils – Ph. D. Dissertation. University of Michigan,

Aubry D., Biarez J., Boelle J.L., Meunier J. (1982). Identification of elastic coefficients through resonant tests of soils samples, *Proceedings 2nd Soils Dynamics and Earthquake Engineering Conference*, Southampton, 65-75.

Ballard R.F., Mac Lean F.G. (1975). Seismic field methods for in situ moduli. *Proceedings of the Conference on Site Measurement of Soil Properties,* ASCE, Special Conference, North Carolina.

Campanella R.G., Stewart W.P. (1992). Seismic cone analysis using digital signal processing for dynamic site characterization, *Canadian Geotechnical Journal*, 29(3).

De Groot D.J., Germaine J.T., Gedney R. (1991). An automated electropneumatic control system for direct simple shear testing, *Geotechnical Testing Journal*, 14(4), 339-348.

De Groot D.J., Germaine J.T., Ladd C.C. (1994). Effect of non uniform stresses on measured DSS stress-strain behaviour, *Journal of Geotechnical Engineering Division*, ASCE, 120(5), 892-912.

Drenvich V.P. (1977). Resonant column testing. Problems and solutions. *Symposium on Dynamic Soil and Rock Testing in the Field and Laboratory for Seismic Studies*, ASTM, Denver.

Duncan J.M., Dunlop P. (1969). Behaviour of soils in simple shear, *Proceedings 7th International Conference on Soil Mechanics and Foundation Engineering*, Mexico, 191-109.

Hardin B.O., Drnevich V.P. (1972). Shear modulus and damping in soils : measurement and parameter effects, *Journal of Soil Mechanics and Foundation Division*, ASCE, vol. 98, SM6.

Haskell N.A. (1953). The dispersion of surface waves on multilayered media, *Bulletin of Seismological Society of America*, vol. 43, n° 1, 17-34.

Hicher P.Y. (1996). Elastic properties of soils, *Journal of Geotechnical Engineering Division*, ASCE, vol. 122, n° 8, 641-648.

Hvorslev M.J., Kaufman R.I. (1952). Torsion shear apparatus and testing procedures, US Waterways Experiment Station, Bulletin 38.

Nigbor R.L., Imai T. (1994) The suspension P-S velocity logging method *in Geophysical Characterization of Sites,* Volume prepared by ISSMFE Technical Committee # 10, R.D. Woods Editor, Balkema.

Prevost J.H., Hoeg K. (1976). Reanalysis of simple shear testing, *Canadian Geotechnical Journal,* vol. 13, n° 4.

Roscoe K.H. (1953). An apparatus for the application of simple shear to soil samples. *Proceedings 3rd International Conference on Soil Mechanics and Foundation Engineering*, London, 186-191.

Silver M.L., Seed H.B. (1971). Deformations characteristics of sands under cyclic loading, *Journal of Soil Mechanics and Foundation Division*, ASCE, vol. 97, n° SM8.

Silver M.L. et al (1976). Cyclic triaxial strength of standard test sand, *Journal of Geotechnical Engineering Division*, ASCE, vol. 102, n° GT5.

Skoglund G.R., Marcuson W.F., Cunny R.W. (1976). Evaluation of resonant column test devices, *Journal of Geotechnical Engineering Division*, ASCE, vol. 102, n° GT11.

Stewart W.P., Campanella R.G. (1993). In situ damping measurement with the seismic cone penetration tests, *Geophysics*.

Stokoe K.H., Hoar R.J. (1978). Variable affecting in situ seismic measurements, *Proceedings of the Conference on Earthquake Engineering and Soil Dynamics*, ASCE, Pasadena.

Stokoe K.H. Wright S.G., Bay J.A., Roesset J.M. (1994). Characterization of geotechnical sites by SASW method *in Geophysical Characterization of Sites,* Volume prepared by ISSMFE Technical Committee # 10, 15-25, R.D. Woods Editor, Balkema.

Sutterer K.G., Frost J.D., Chameau J.L.A. (1996). Polymer impregnation to assist undisturbed sampling of cohesionless soils, *Journal of Geotechnical Engineering Division*, ASCE, vol. 122, n° 3, 209-215.

Thomson W.T. (1950). Transmission of elastic waves through a stratified soil medium, *Journal of Applied Physics*, vol. 21, 89-93.

Vucetic M., Lacasse S. (1982). Specimen size effect in simple shear tests, *Journal of Geotechnical Engineering Division*, ASCE, 108(2), 1567-1585.

Whiteley R.J. (1994). Seismic refraction testing – A tutorial, *in Geophysical Characterization of Sites*- Volume prepared by ISSMFE Technical Committee # 10, 45-47, R.D. Woods Editor, Balkema.

Woods R.D. (1978). Measurement of dynamic soil properties, *Proceedings of the Conference on Earthquake Engineering and Soil Dynamics*, ASCE, Pasadena.

Woods R.D. (1994). Geophysical characterization of sites, Volume prepared by ISSMFE Technical Committee # 10, Balkema.

Standards

Drnevich V.P., Hardin B.O., Shippy D.J. – Modulus and damping of soils by the resonant column method – Dynamic Geotechnical Testing, ASTM SPT654, pp. 91-125, 1978.

Standard Test Methods for Cross-hole Seismic Testing – ASTM D4428/D4428M-84, 1984.

Standard Method for deep quasi-static, cone and friction cone penetration tests of soils. ASTM-D3441, American Society of Testing Materials, 1979.

Soil Structure Interaction

Alain Pecker[1,2]

[1] Solid Mechanics Laboratory, CNRS UMR 7649, Department of Mechanics, Ecole Polytechnique, Palaiseau, France
[2] Géodynamique et Structure, Bagneux, France

Abstract. Usually in the seismic design of ordinary building, soil structure interaction is neglected and the dynamic response of the structure is evaluated under the assumption of a fixed based response. However during seismic loading the soil undergoes deformations which are imposed to the foundation; the question naturally arises of knowing if the motion in the vicinity of the structure is altered by the presence of the structure and how the structure response is modified by the compliance of the supporting soil. This interaction between the structure and the soil is named soil-structure interaction (SSI). The purpose of this chapter is to illustrate whether and under which conditions SSI is important and what are its consequences on the dynamic response of the structure.

1 Fundamentals of Soil Structure Interaction

The earthquake design loads applied to the foundation arise from the inertia forces developed in the superstructure and from the soil deformations, caused by the passage of seismic waves, imposed on the foundations. These two phenomena are referred in the technical literature as inertial and kinematic loading. The relative importance of each factor depends on the foundation characteristics and nature of the incoming wave field.

The generic term encompassing both phenomena is Soil-Structure Interaction (SSI). However, more often, design engineers refer to inertial loading as SSI, ignoring the kinematic component. This situation stems from the fact that:

- kinematic interaction may in some situations be neglected;
- aseismic building codes, except for very few exceptions like Eurocode 8, do not even mention it;
- kinematic interaction effects are far more difficult to evaluate rigorously than inertial interaction effects.

Figure 1 illustrates the key features of the problem under study (Gazetas-Mylonakis, 1998). It is presented in the general situation of an embedded foundation supported on piles but all the conclusions are valid for any foundation type. The soil layers away from the structure are subjected to seismic excitation consisting of numerous incident waves: shear waves (S waves), dilatational waves (P wave), surface waves (R or L waves). The nature of the incoming waves is dictated by seismological conditions but the geometry, stiffness and damping characteristics of the soil deposit modify this motion; this modified motion is the free field motion at the site of the foundation. Determination of the free field motion is in itself a challenging task because, as pointed out by

Lysmer (1978), the design motion is usually specified at only one location, the ground surface, and the complete wave field cannot be back-calculated from this incomplete information; that is the problem is mathematically ill posed. Assumptions have to be made regarding the exact composition of the free field motion and it can be stated that no satisfactory solution is available to date.

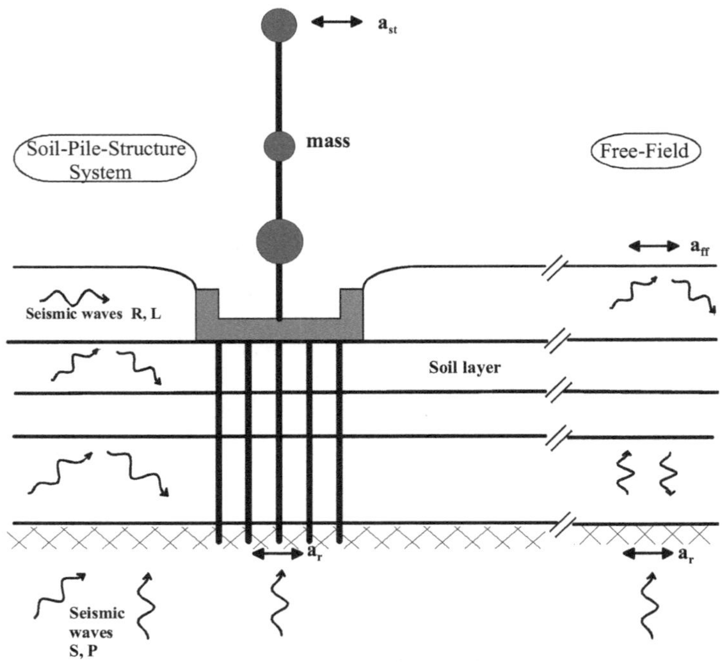

Figure 1 : Illustration of soil structure interaction on the structure response

Let us now consider the motion around the structure and its foundation: the seismically deforming soil will force the piles and the embedded foundation to move, and subsequently the supported structure. Even without the superstructure, the motion of the foundation will be different from the free field motion because of the differences in rigidity between the soil on the one hand, and the piles and foundations on the other hand; the incident waves are reflected and scattered by the foundation and piles which in turn are stressed developing curvatures and bending moments. This is the phenomenon of kinematic interaction. The motion induced at the foundation level generates oscillations in the superstructure which develop inertia forces and overturning moments at its base. Thus the foundation, the piles, and eventually the surrounding soil experience additional dynamic forces and displacements. This is the phenomenon of inertial interaction.

Obviously the foundation, in a broad sense, must be checked for the combined inertial and kinematic loading.

For the evaluation of SSI effects of linear systems the most appropriate constitutive model for the soil is either the linear elastic or, more commonly, the equivalent viscoelastic linear model.

However, as it will be shown, soil non linearities can often be accounted for in an approximate manner, just by choosing appropriate values for the soil parameters.

2 Illustration of SSI on a simple example

The effect of soil structure interaction can be illustrated with the simple idealized model of the structure depicted in Figure 2 (Wolf, 1985).

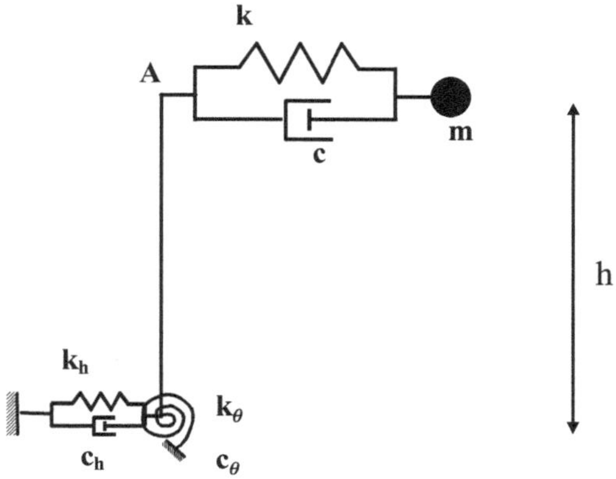

Figure 2: Idealized model for SSI illustration

The structure is modeled by a mass, a spring and a dashpot placed at an elevation h above the foundation. The connection between the structure and the foundation is ensured by a rigid beam. The foundation rests on the soil deposit and its interaction with the soil is, for the time being, modeled by springs and dashpots, called foundation impedances that will be defined later. The spring represents the stiffness of the supporting medium and the dashpot reflects the dissipation of energy arising from the soil itself (material damping) and from the radiation of the seismic waves away from the foundation. For the sake of simplicity, material damping is neglected with respect to radiation damping, which is a valid approximation for homogeneous soil deposits at moderate strain amplitudes. The system depicted in Figure 2 possesses 3 degrees of freedom:

- The horizontal displacement of the mass u_m
- The horizontal displacement of the foundation u_0
- The rotation θ of the foundation

It is subjected to a horizontal harmonic support displacement with circular frequency ω and amplitude u_g. The dynamic equilibrium equations of the system can be easily established from Lagrange equations with the following generalized coordinates

- $q_1 = u$ the relative displacement of the mass with respect to A
- $q_2 = u_0$ the horizontal displacement of the foundation
- $q_3 = \theta$ the rotation of the foundation

The absolute displacement u^t of the mass is related to the previous parameters by :

$$u^t = u_g + u_0 + u + h\theta \tag{1}$$

The kinetic energy of the system is given by

$$T = \frac{1}{2} m\left(\dot{u}_g + \dot{u}_0 + \dot{u} + h\dot{\theta}\right)^2 \tag{2}$$

And its potential energy by:

$$V = \frac{1}{2}\left(ku^2 + k_h u_0{}^2 + k_\theta \theta^2\right) \tag{3}$$

The work of the non conservative forces is coming from the dashpots:

$$\delta W = -\left(C\,\dot{u}\delta u + C_h\,\dot{u}_0\delta u_0 + C_\theta\,\dot{\theta}\,\delta\theta\right) \tag{4}$$

Introducing the dimensionless damping ratios

$$\xi = \frac{\omega C}{2k}, \qquad \xi_h = \frac{\omega C_h}{2k_h} \qquad \xi_\theta = \frac{\omega C_\theta}{2k_\theta} \tag{5}$$

the dynamic equilibrium equations in the frequency domain are:

$$\begin{cases} -m\,\omega^2(u_0 + u + h\theta) + k(1 + 2i\,\xi)u = m\,\omega^2\,u_g \\ -m\,\omega^2(u_0 + u + h\theta) + k_h(1 + 2i\,\xi_h)u_0 = m\,\omega^2\,u_g \\ -m\,h\,\omega^2(u_0 + u + h\theta) + k_\theta(1 + 2i\,\xi_\theta)\theta = m\,h\,\omega^2\,u_g \end{cases} \tag{6}$$

An elimination of the variables u_0 and θ between the three equations (6), introducing the notations $m\omega_s^2 = k$, $m\omega_h^2 = k_h$, $mh^2\omega_\theta^2 = k_\theta$ given that ξ, ξ_h, $\xi_\theta \ll 1$ leads to the relative displacement of the mass:

$$\left[1 + 2i\xi - \frac{\omega^2}{\omega_s^2} - \frac{\omega^2}{\omega_h^2}(1 + 2i\xi - 2i\xi_h) - \frac{\omega^2}{\omega_\theta^2}(1 + 2i\xi - 2i\xi_\theta)\right]u = \frac{\omega^2}{\omega_s^2}u_g \tag{7}$$

The response of a one degree of freedom oscillator with the same mass m, an eigenfrequency $\tilde{\omega}$ and a damping ratio $\tilde{\xi}$ subjected to a harmonic displacement with amplitude $\tilde{u}_g$ and frequency ω at its base is equal to:

$$\left(1 + 2i\tilde{\xi} - \frac{\omega^2}{\tilde{\omega}^2}\right)u = \frac{\omega^2}{\tilde{\omega}^2}\tilde{u}_g \tag{8}$$

The one degree of freedom oscillator will have the same response as the SSI system provided the following relations hold:

$$\frac{1}{\tilde{\omega}^2} = \frac{1}{\omega_s^2} + \frac{1}{\omega_h^2} + \frac{1}{\omega_\theta^2}$$
$$\tilde{\xi} = \frac{\tilde{\omega}^2}{\omega_s^2}\xi + \frac{\tilde{\omega}^2}{\omega_h^2}\xi_h + \frac{\tilde{\omega}^2}{\omega_\theta^2}\xi_\vartheta \tag{9}$$
$$\tilde{u}_g = \frac{\tilde{\omega}^2}{\omega_s^2}u_g$$

Examination of equations (9) shows that the effect of soil structure interaction is

- To decrease the frequency of vibration of the fixed base structure ($\tilde{\omega} < \omega_s$);
- To increase the damping ratio of the system ($\tilde{\xi} > \xi$) with respect to the fixed base structure;
- To decrease the amplitude of the effective input motion at the base of the structure ($\tilde{u}_g < u_g$).

The previous conclusions are illustrated in Figure 3 which presents for a circular foundation resting on the surface of a homogeneous elastic halfspace the variations of the quantities $\tilde{\omega}/\omega_s$, $\tilde{\xi}$, $\tilde{u}_g/u_g$ as a function of the dimensionless parameters

$$\bar{h} = \frac{h}{r} \quad , \quad s = \frac{\omega_s h}{V_s} \quad , \quad \bar{m} = \frac{m}{\rho r^3} \tag{10}$$

Whre r is the radius of the foundation, V_s and ρ the shear wave velocity and mass density of the halfspace. Figure 3 clearly shows that soil structure interaction is more pronounced for soft soil conditions (increasing s) and for heavy structures (increasing m).

3 Formulation of a Soil Structure Interaction Problem

Before examining the different ways to take into account SSI in a dynamic analysis, it is useful, and illustrative, to formulate the problem in a general sense. That formulation is presented within the framework of the finite element method. In fact the complexity of the problem to solve is beyond the capability of closed form solutions and numerical solutions are required. However, other numerical techniques can be used, such as boundary element techniques; nevertheless, the concepts that are presented below are general and not restricted to finite element solutions; as a matter of fact the results could also have been obtained from the Principle of Virtual Rate of Work. The dynamic equilibrium equations are obtained with reference to Figure 4, which is a schematic representation of a SSI problem.

Denoting $[\mathbf{M}]$, $[\mathbf{C}]$ and $[\mathbf{K}]$ the mass, damping and stiffness matrices the dynamic equilibrium equations are:

$$[\mathbf{M}]\{\ddot{\mathbf{u}}\} + [\mathbf{C}]\{\dot{\mathbf{u}}\} + [\mathbf{K}]\{\mathbf{u}\} = \{\mathbf{Q}_f\} \tag{11}$$

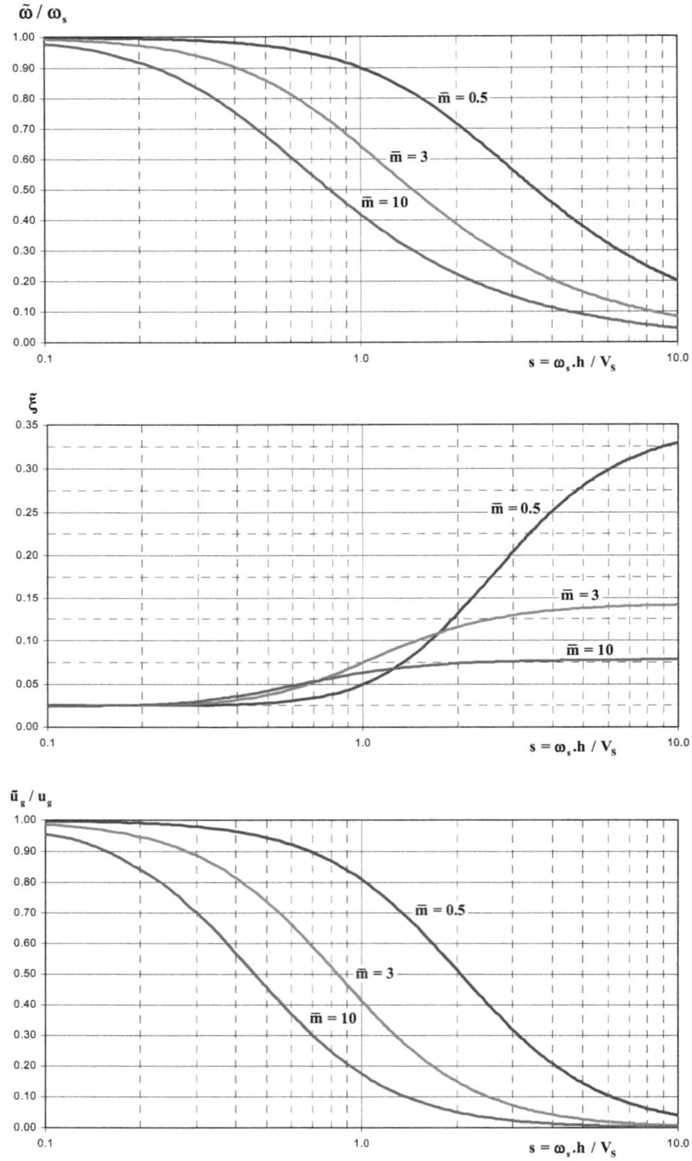

Figure 3 : Illustration of the influence of soil structure interaction

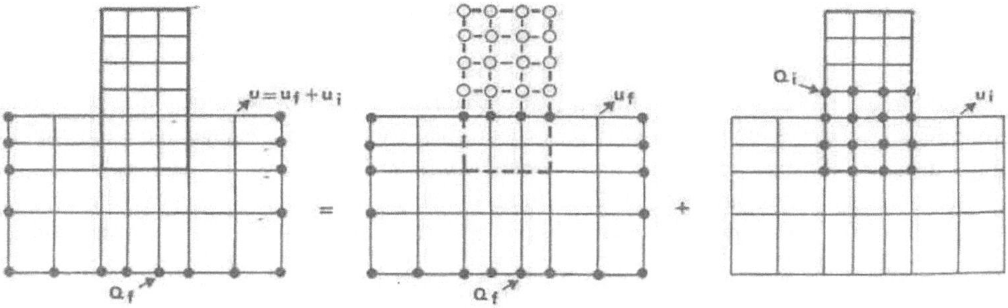

Figure 4 : Decomposition of a SSI problem

The source of the incident motion (earthquake focus) is usually not included within the finite element model; therefore, the loading vector $\{\mathbf{Q}_f\}$ has non zero values on the boundary of the model. This boundary is assumed to be sufficiently remote from the structure in order that the motion at the boundary is not influenced by the presence of the structure. Without the structure the equations of motion are identical to equations (11) with indices "f" representing the quantities related to the free field soil:

$$[\mathbf{M}_f]\{\ddot{\mathbf{u}}_f\}+[\mathbf{C}_f]\{\dot{\mathbf{u}}_f\}+[\mathbf{K}_f]\{\mathbf{u}_f\}=\{\mathbf{Q}_f\} \tag{12}$$

It is worth noting that equation (12) can only be solved provided some hypotheses are made with respect to the nature and direction of the incident motion. Introducing the interaction displacement $\{\mathbf{u}_i\}$ defined by:

$$\{\mathbf{u}\}=\{\mathbf{u}_i\}+\{\mathbf{u}_f\} \tag{13}$$

The equation satisfied by $\{\mathbf{u}_i\}$ is

$$[\mathbf{M}]\{\ddot{\mathbf{u}}_i\}+[\mathbf{C}]\{\dot{\mathbf{u}}_i\}+[\mathbf{K}]\{\mathbf{u}_i\}=-\{\mathbf{Q}_i\} \tag{14}$$

with

$$\{\mathbf{Q}_i\}=[[\mathbf{M}]-[\mathbf{M}_f]]\{\ddot{\mathbf{u}}_f\}+[[\mathbf{C}]-[\mathbf{C}_f]]\{\dot{\mathbf{u}}_f\}+[[\mathbf{K}]-[\mathbf{K}_f]]\{\mathbf{u}_f\} \tag{15}$$

The load vector $\{\mathbf{Q}_i\}$ is determined from the free field displacements. For linear systems equations (13) to (15) show that the SSI problem is decomposed in the sum of a site free field response analysis (equation (12)) and of a source problem in which the load vector $\{\mathbf{Q}_i\}$ has only non zero components at the nodes common to the structure and to the soil (basement) as illustrated in Figure 4. The total displacement for the SSI problem is then given by equation (13).

Equation (15) clearly points out that there is interaction as soon as there is a difference in stiffness and/or mass between the soil and the structure. To simplify the demonstration let us leave out the damping term in equation (15) and restrict our problem to that of a structure at the surface of a horizontally layered soil profile subjected to the vertical propagation of body waves. Under these conditions, in the free field, all the points at the soil surface move in phase; if the foundation is infinitely stiff the last term in equation (15) vanishes; the load vector reduces to:

$$\{\mathbf{Q}_i\} = \left[[\mathbf{M}] - [\mathbf{M}_f]\right]\{\ddot{\mathbf{u}}_f\} \tag{16}$$

The forces $\{\mathbf{Q}_i\}$ at the base of the structure give rise to a support motion, equivalent to an inertial force field in the superstructure. Consequently, interaction is only generated by inertial forces in the structure; this phenomenon is named *inertial interaction.*

Let us consider now an embedded structure, the mass of which is zero above the ground and equal and equally distributed to the soil mass for the embedded part. The load vector becomes:

$$\{\mathbf{Q}_i\} = \left[[\mathbf{K}] - [\mathbf{K}_f]\right]\{\mathbf{u}_f\} \tag{17}$$

These forces arise only from the difference in stiffness for the embedded part between the structure and the soil. Even with the same mass, there is interaction; this kind of interaction is named *kinematic interaction.* It arises from the stiffness of the foundation that prevents it from following the displacements imposed by the soil. That kind of interaction may be equal to zero as shown previously for surficial foundations, or negligible under certain circumstances, like very flexible piled foundations. However, for stiff embedded structures it may be very significant.

In the most general situation, soil structure interaction arises from both phenomena: inertial and kinematic interaction. Figure 4 and the previous developments illustrate the two broad approaches for evaluating SSI; Figure 4(a) corresponds to the direct methods the solution of which is obtained by a direct solution of equation (11). This approach does not involve any superposition and is therefore well suited for non linear systems. Alternatively, substructure methods take advantage of the decomposition of Figure 4(b) and (c), or of similar decompositions, to solve the global problem in successive steps. These methods are obviously only applicable to linear problems.

4 Superposition Theorem for Soil Structure Interaction

Decomposition of SSI in inertial interaction and kinematic interaction, as exposed in paragraph 3 is not only convenient to illustrate the fundamentals of SSI but also give rise to solution techniques based on sub-structuration methods. The validity of these methods relies on the superposition theorem established by Kausel (1978) and Roesset (1973). This theorem states that the response of the system depicted in Figure 4(a), subjected to a base acceleration $\ddot{u}_g$ can be obtained by:

- Either with a direct solution of equation $[\mathbf{M}]\{\ddot{\mathbf{u}}\} + [\mathbf{K}]\{\mathbf{u}\} = -[\mathbf{M}]\{\mathbf{I}\}\,\ddot{u}_g$ (18)

with

$\{\mathbf{u}\}$	the vector of relative displacement with respect to the base f the model
$\{\mathbf{I}\}$	a unit vector giving the direction of load application
$[\mathbf{M}]$ and $[\mathbf{K}]$	the mass and stiffness matrices

- or in two steps by defining the vector of relative displacement $\{\mathbf{u}\}$ as the sum of kinematic interaction displacement $\{\mathbf{u}_{kin}\}$ and of inertial interaction displacements $\{\mathbf{u}_{iner}\}$:

$$\{\mathbf{u}\} = \{\mathbf{u}_{\text{kin}}\} + \{\mathbf{u}_{\text{iner}}\} \tag{19}$$

and by solving the simultaneous system of differential equations

$$[\mathbf{M}_{\text{Soil}}]\{\ddot{\mathbf{u}}_{\text{kin}}\} + [\mathbf{K}]\{\mathbf{u}_{\text{kin}}\} = -[\mathbf{M}_{\text{Soil}}]\{\mathbf{I}\}\ddot{u}_g \tag{20}$$

$$[\mathbf{M}]\{\ddot{\mathbf{u}}_{\text{iner}}\} + [\mathbf{K}]\{\mathbf{u}_{\text{iner}}\} = -[\mathbf{M}_{\text{St}}]\left[\{\ddot{\mathbf{u}}_{\text{kin}}\} + \{\mathbf{I}\}\ddot{u}_g\right] \tag{21}$$

in which $[\mathbf{M}_{\text{Soil}}]$ and $[\mathbf{M}_{\text{St}}]$ represent the mass matrices of the soil substructure and of the structure $[\mathbf{M}] = [\mathbf{M}_{\text{Soil}}] + [\mathbf{M}_{\text{St}}]$.

Equivalence between equation (18) and equations (20)-(21) is established by simple addition taking into account equation (19) and the previous definition of the mass matrices.

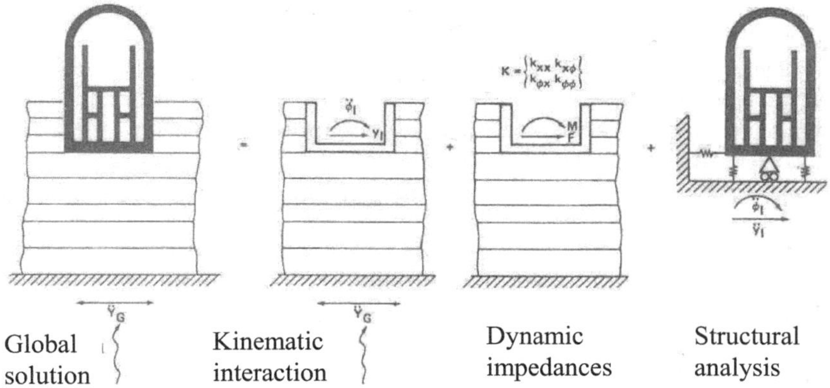

Figure 5 : Superposition theorem (Kausel, 1978)

Equation (20) gives the response of a massless structure to the incident motion $\ddot{u}_g$. Its solution provides the kinematic interaction motions that are used as input motions for the solution of equation (21). In the solution of equation (21) the soil can be modeled with finite elements or equivalently by a stiffness matrix representing the condensation of all the foundation-soil degrees of freedom at the interface; this condensation is only possible in the frequency domain. In that framework the stiffness matrix is formed with the complex valued moduli taking into account material damping. The stiffness matrix is composed of a real part (the stiffness) and of an imaginary part representing the energy dissipation arising either from material damping and/or from radiation damping. The terms in the matrix are frequency dependent.

For a rigid foundation, it is legitimate to replace the (N x N) stiffness matrix (N being the number of degrees of freedom at the interface) by a (6 x 6) matrix providing the rigid body motions of the foundation. This matrix is called the *impedance matrix* and can be conceptually viewed as an assemblage of springs and dashpots. It follows that the kinematic interaction motions are the rigid body motions of the massless structure.

The structure of equation (21) reveals that the solution $\{\mathbf{u}_{\text{iner}}\}$ can be interpreted as the vector of relative displacements with respect to a fictitious support subjected to the rigid body motions of the foundation. Therefore, under the assumption of a rigid foundation, it is pertinent to split the global problem into there sub-problems:

- determination of the motion of the massless rigid foundation subjected to the seismic design motion; this steps represents the solution of equation (20);
- determination of the foundation impedance matrix; this matrix is composed of a real and an imaginary component, both being frequency dependent;
- calculation of the dynamic response of the structure connected to the foundation impedances and subjected, at its support, to the kinematic interaction motions.

As long as the foundation is perfectly rigid, this three steps approach is strictly equivalent to the resolution of the global problem (equation (18)). The advantage of this decomposition is obvious if one of the successive steps can be simplified or ignored: the first step always exists except for a surficial foundation resting at the surface of a horizontal layered soil profile subjected to the upward propagation of body waves; in the latter situation solution to step 1 is identical to solving the free field site response since kinematic interaction is nil. Solution to the second step can be simplified for common geometries by using published results in the literature. The third step is always required; however it is simpler and more common to structural engineers since it resorts to classical dynamic analyses.

References

Gazetas G., Mylonakis G. (1998). Seismic soil structure interaction : new evidence and emerging issues, *Geotechnical Earthquake Engineering and Soil Dynamics*, ASCE, II, 1119-1174.

Kausel E.,(1974). Forced vibrations of Circular Foundations on Layered Media, Soils Publication n° 336, Department of Civil Engineering, Massachusetts Institute of Technology, Cambridge.

Kausel E., Whitman A., Murray J., Elsabee F.,(1978). The Spring Method for Embedded Foundations, *Nuclear Engineering and Design*, Vol. n° 48.

Lysmer J.,(1978). Analytical Procedures in Soil Dynamics, *State of the Art, ASCE Conference on Soil Dynamics and Earthquake Engineering*, Pasadena, California.

Pecker, A.(1984). *Dynamique des sols*, Presses de l'Ecole Nationale des Ponts et Chaussées.

Roesset J.M., Whitman R.V., Dobry R., (1973). Modal Analysis for Structures with Foundation Interaction, *Journal of Structural Engineering Division,* ASCE, Vol. 99 n° ST3, 89-416.

Wolf J.P.,(1985). Dynamic Soil Structure Interaction, Prentice Hall Englewood Cliffs, New Jersey.

Earthquake Foundation Design

Alain Pecker[1,2]

[1] Solid Mechanics Laboratory, CNRS UMR 7649, Department of Mechanics, Ecole Polytechnique, Palaiseau, France
[2] Géodynamique et Structure, Bagneux, France

Abstract. Earthquake foundation design is a challenging task that requires analytical capabilities and extensive understanding of soil behaviour and soil structure interaction. The classical approach involves the determination of the forces applied to the foundation, the seismic demand, and the verification of the bearing capacity, the seismic capacity. However not all situations can be tackled with analyses. Seismic building codes and in particular their chapters on foundation detailing are fundamental to achieve a safe design.

1 Overview of Foundation Design

Design of foundations still remains a challenging task for the earthquake geotechnical engineer. Leaving aside the seismic retrofit of existing foundations, which is an even more difficult issue, the design of new foundations raises issues which are far from being totally resolved. One of the main reasons stems from the complexity of the problem which requires skills in soil mechanics, foundation engineering, and soil-structure interaction along with, at least, some knowledge of structural dynamics.

A parallel between static design and seismic design reveals some similarity but also very marked differences. In the early days, static design of foundations put much emphasis on the so-called bearing capacity problem (failure behavior); with the introduction of an appropriate safety factor, close to 3, the short term settlements were deemed to be acceptable for the structure. It is only with the increase in the understanding of soil behavior and the development of reliable constitutive models that sound predictions of settlements could be achieved. Not surprisingly, earthquake geotechnical engineers have focused their attention on the cyclic non linear behavior of soils and on the evaluation of the cyclic deformations of foundations. This was clearly dictated by the need for an accurate evaluation of the soil-structure interaction forces which govern the structural response. It is only during the last decade that seismic bearing capacity problems and evaluation of permanent displacements have been tackled. These studies have clearly been motivated by the foundation failures observed in the Mexico City (1985) and Kobe (1995) earthquakes.

These two aspects of foundation design have reached a state of development where they can be incorporated in seismic building codes; Eurocode 8 - Part 5 is certainly a pioneering code in that respect. Nevertheless, a new trend is emerging in earthquake engineering, known as "Performance Based Design" (PBD), which definitely needs to be accounted for in earthquake foundation engineering. In this lecture we will focus only on the evaluation of the seismic demand and seismic capacity and review the code approach, and foundation detailing, the earthquake resistant desin of foundations.

2 Aseismic Design Process

The aseismic design process for foundations is a "very broad activity requiring the synthesis of insight, creativity, technical knowledge and experience" (Pender, 1995). Information is required and decisions have to be made at various stages including (Pecker and Pender, 2000):

1. the geological environment and geotechnical characterization of the soil profile;
2. the investigation of possible solutions
3. the definition of the loads that will be applied to the foundation soil by the facility to be constructed;
4. information about the required performance of the structure;
5. the evaluation of load capacity, assessment of safety factors and estimates of deformations;
6. consideration of construction methods and constraints that need to be satisfied (finance and time);
7. exercise of judgment to assess potential risks.

Obviously the process described above is not a linear progression. Several iterations may be required, at least from step (2) to step (7), before arriving at a feasible, reliable and economic design. In the following we will focus on steps (3) and (5). We will assume that all the required information related to the soil characterization and structural performance is available. This in no way means that these two items are of secondary importance; the data listed under these items are probably the most difficult to assess and considerable experience is required as well as the exercise of judgment.

3 Code Approach to Foundation Design

In almost every seismic building code, the structure response and foundation loads are computed neglecting soil-structure interaction; the dynamic response is obtained from a fixed base analysis of the structure. The belief is that SSI always plays a favorable role in decreasing the inertia forces; this is clearly related to the standard shape of code spectra which almost invariably possess a gently descending branch beyond a constant spectral acceleration plateau. Lengthening of the period, due to SSI, moves the response to a region of smaller spectral accelerations (Figure 1). However there is evidence that some structures founded on unusual soils are vulnerable to SSI. Examples are given by Gazetas and Mylonakis (1998) for instance.

This has been recognized in some codes. Eurocode 8 states that:

"The effects of dynamic soil-structure interaction shall be taken into account in the case of:

- structures where P-δ effects play a significant role;
- structures with massive or deep seated foundations;
- slender tall structures;
- structures supported on very soft soils, with average shear wave velocity less than 100 m/s.

The effects of soil-structure interaction on piles shall be assessed..."

In addition, an annex to the code describes the general effect of SSI and a specific chapter analyzes its effects on piles and the way to deal with it. To the best of our knowledge Eurocode 8 is the only code which recognizes the importance of kinematic interaction for piled foundations:

"Bending moments developing due to kinematic interaction shall be computed only when two or more of the following conditions occur simultaneously :

- the subsoil profile is of class C (soft soil), or worse, and contains consecutive layers with sharply differing stiffness,
- the zone is of moderate or high seismicity, $\alpha > 0.10$,
- the supported structure is of importance category III or IV."

Note that implicitly for "normal" soil profiles and ordinary buildings kinematic interaction need not be computed.

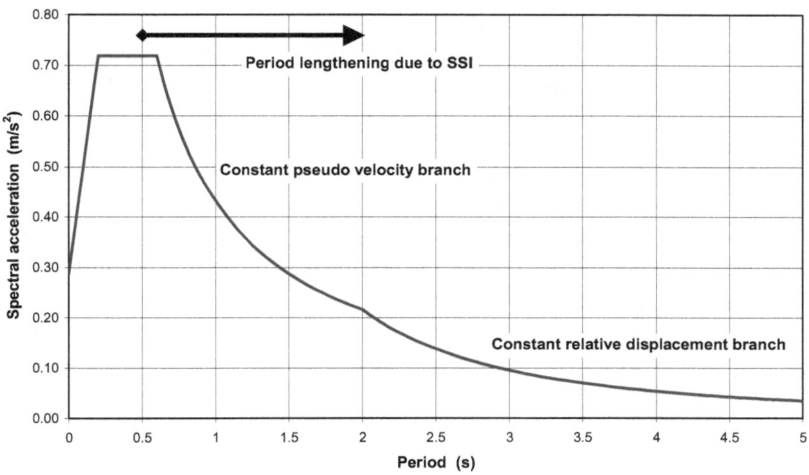

Figure 1: Standard spectral shape

4 Evaluation of Seismic Demand

4.1 Linear systems

The standard practice in earthquake foundation design is, on the one hand, to evaluate the seismic forces acting at the foundation (*seismic demand*) and, on the other hand, to check the foundation capacity (*seismic bearing capacity*).

With the tremendous development of computer facilities, there does not seem to be any rational reason for neglecting soil-structure interaction. Most building codes now require that the structural response be evaluated using a multimodal analysis, as opposed to a former monomodal analysis and this can be performed with most computer codes available on the market. Referring to the multistep approach (Kausel, 1978), the last step of an SSI analysis (response of the structure connected to the impedances) can be performed on a routine basis provided that:

- the system remains linear;
- kinematic interaction can be neglected;
- dynamic impedance functions are readily available.

Although the superposition theorem is exact for linear soil, pile and structure, it can nevertheless be applied to moderately non linear systems. This can be achieved by choosing reduced soil characteristics which are compatible with the free field strains induced by the propagating seismic waves: this is the basis of the equivalent linear method, pioneered by Idriss and Seed (1968). This engineering approximation implies that all the soil non linearities arise from the passage of the seismic waves and that additional non linearities, developed around the edges of a mat foundation or along the piles shafts, are negligible. Experience shows that it is a valid approximation in many situations where large soil instabilities do not occur.

For some situations, kinematic interaction can be neglected and the second step of the multistep approach can be bypassed. It must be realized however that, if kinematic interaction is thought to be significant, there is no simple means for evaluating it; as a matter of fact, evaluation of kinematic interaction is almost as difficult as solving the complete SSI problem. Obviously kinematic interaction is exactly zero for shallow foundations in a seismic environment consisting exclusively of vertically propagating shear waves or dilatational waves. Gazetas (1984) has demonstrated that when the piles are flexible with respect to the surrounding soil, kinematic interaction is significant for small to medium frequencies. During the last decade, numerous solutions for the dynamic impedances of any shape foundations and of piles have been published (Gazetas, 1990). They are available for homogeneous soil deposits but also for moderately heterogeneous ones. In addition, simplified methods are available in the case of pile foundations to account for the group effect (Dobry and Gazetas, 1988).

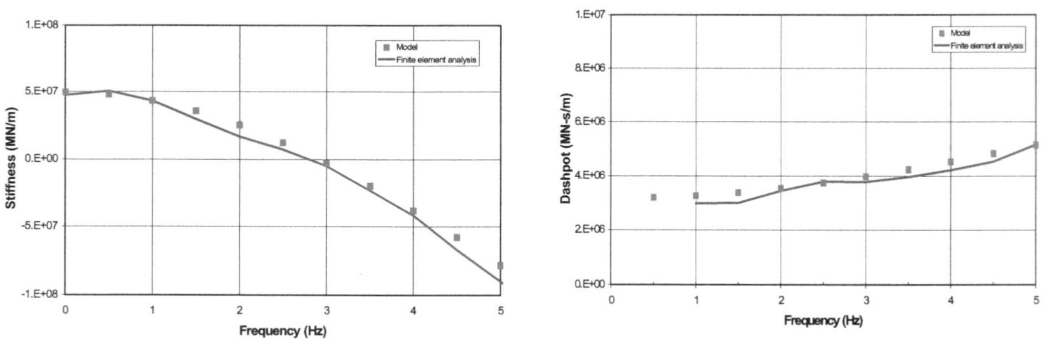

Figure 2 : Rocking dynamic impedances – example

Therefore provided all the aspects listed above are properly covered, seismic soil structure interaction can be dealt with at a minimal cost and reduces to the last step of the multistep approach: dynamic response of the structure connected to the impedance functions and subjected to the free field motion (equal to the kinematic interaction motion). However to be fully efficient, and to allow for the use of conventional dynamic computer codes, the impedance functions which are frequency dependent (Figure 2) must be represented by frequency independent values. The simplest version of these frequency independent parameters is the so-called springs and dashpots assembly. From the published results, it appears that only under very restrictive soil conditions (homogeneous halfspace, regular foundations) can these dynamic impedances be represented by constant springs and dashpots. Nevertheless, structural engineers still proceed using these values which, more than often, are evaluated as the static component (zero frequency) of the impedance functions.

However, fairly simple rheological models can be used to properly account for the frequency dependence of the impedance functions. These models can be developed using curve fitting techniques, or with physical insight, such as the series of cone models developed by Wolf (1994). Figure 3 shows examples of such models: Figure 3a is the model proposed by De Barros and Luco (1990) based on a curve fitting technique; Figure 3b is a class of cone models proposed by Wolf. With such models, which are most conveniently used in time history analyses, the actual dynamic action of the soil can be properly accounted for; even "negative stiffnesses", which are frequently encountered in layered soil profiles, can be apprehended with those models. As an illustrative example, Figure 2 presents the application of model 3a to an actual bridge pier foundation; the foundation is a large circular caisson, 90 m in diameter, resting on a highly heterogeneous soft soil profile. The "exact" impedances were computed using a frequency domain finite element analysis. Note the very good fit achieved by the model (square symbols) even for the negative stiffness of the rocking component. Clearly, implementation of such simple rheological models does not impose a heavy burden to the analyst and represents a significant improvement upon the lengthy and tedious iteration process in which springs and dashpots are updated to become compatible with the SSI frequencies.

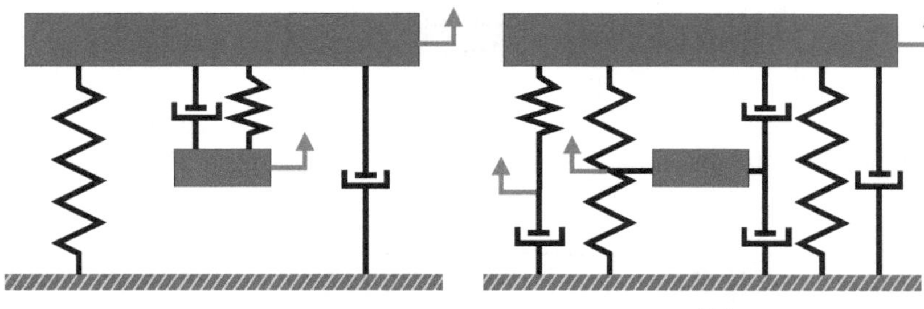

Figure 3 : Example of cone models

Before moving to consideration of nonlinear SSI, there in one section of Eurocode 8 which provides for a transition between the linear elastic approach discussed above and non linear methods discussed below. Table 1 (taken from Eurocode 8) acknowledges that with increasing ground acceleration the soil adjacent to a shallow foundation will experience increasing shear strains and consequently the stiffness will decrease and the material damping increase. Table 1 suggests how the apparent average shear modulus and material damping of the soil adjacent will change with increasing peak ground acceleration and envisages that an elastic SSI calculation would be done with the modified values for the soil stiffness and damping. (Following this simplification there is, of course, no frequency dependence on the stiffness and damping parameters for the foundation).

Table 1: Average soil damping factors and average reduction factors (± one standard deviation) for shear wave velocity v_s and shear modulus G within 20 m depth.($v_{s\,max}$ = average v_s value at small strain (< 10^{-5}), not exceeding 300 m/s. G_{max} = average shear modulus at small strain.)

Ground acceleration ratio, α	Damping factor	$\frac{v_s}{v_{s,maz}}$	$\frac{G}{G_{max}}$
0,10	0,03	0,9(±0,07)	0,80(± 0,10)
0,20	0,06	0,7(± 0,15)	0,50(+ 0,20)
0,30	0,10	0,6(+0,15)	0,35(± 0,20)

4.2 Non linear systems

One of the main limitations of the multistep approach is the assumption of linearity of the system for the superposition theorem to be valid. As noted previously, some non linearities, such as those related to the propagation of the seismic waves, can be introduced but the non linearities specifically arising from soil-structure interaction are ignored. The generic term "non linearities" covers geometrical non linearities, such as foundation uplift, and material non linearities, such as soil yielding around the edges of shallow foundations, along the shafts of piles, and the formation of gaps adjacent to pile shafts. Those non linearities may be beneficial and tend to reduce the forces transmitted by the foundation to the soil and therefore decrease the seismic demand. This has long been recognized for foundation uplift for instance (see ATC 40).

Giving up the mathematical rigor of the superposition theorem, an engineering approximation to these aspects can be reached by substructuring the supporting medium into two sub-domains (Figure 4).

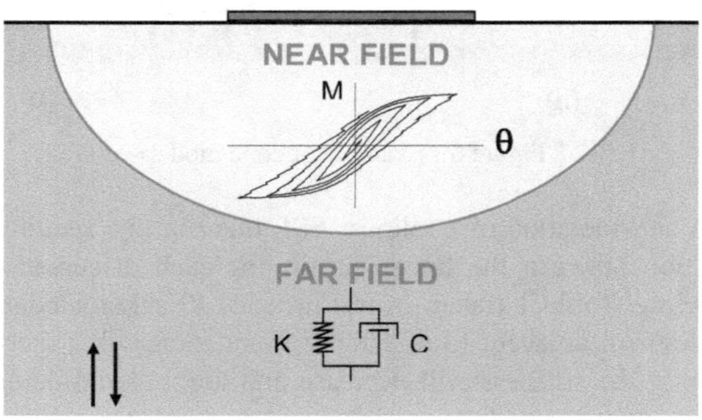

Figure 4 : Conceptual subdomains for dynamic soil structure analyses

- a far field domain, which extends a sufficient distance from the foundation for the soil structure interaction non linearities to be negligible; non linearities in that domain are only governed by the propagation of the seismic waves,
- a near field domain, in the vicinity of the foundation where all the geometrical and material non linearities due to soil structure interaction are concentrated.

The exact boundary between both domains is not precisely known but its location is irrelevant for practical purposes. This concept of far field and near field domains can be easily implemented if one assumes that the degrees of freedom of the foundation are uncoupled: the far field domain is modeled with the linear (or equivalent linear) impedance functions whereas the near field domain is lumped into a non-linear macro-element. A simplified rheological representation of this sub-structuring is shown in Figure 5 (Pecker, 1998): the macro-element is composed of a finite number of springs and Coulomb sliders which are determined from curve fitting to the non-linear force-displacement (or moment-rotation) backbone curve, computed for instance with a static finite element analysis (push over analysis).

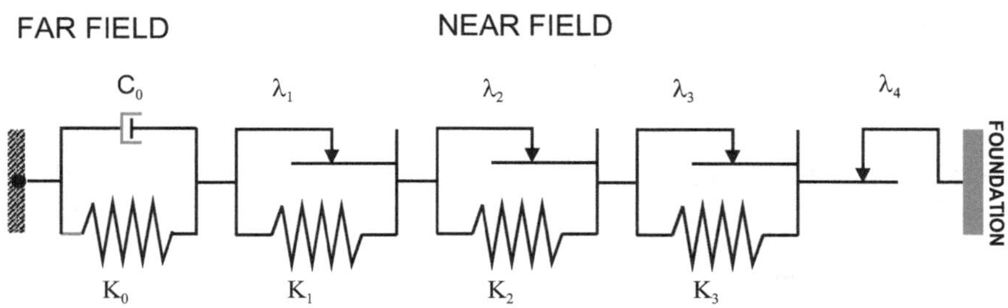

Figure 5 : Non linear rheological model for SSI

Damping in the near field domain arises only from material damping and obeys Masing's law; damping in the far field domain is of the viscous type. Calibration of this simplified rheological model against a rigorous 2D dynamic finite element analysis, including all the non linearities mentioned previously, shows very promising results. This model can be extended in a more rigorous way to account for the coupling between the various degrees of freedom of the foundation, especially between the vertical and rotational ones when uplift occurs (Cremer et al, 2000).

5 Evaluation of Seismic Capacity : Foundation Bearing Capacity

Once the forces transmitted to the soil by the foundation are determined, the design engineer must check that these forces can be safely supported: the foundation must not experience a bearing capacity failure or excessive permanent displacements. At this point a major difference appears between static, permanently acting loads, and seismic loads. In the first instance excessive loads generate a general foundation failure whereas seismic loads, which by nature vary in time, may induce only permanent irrecoverable displacements. Therefore, failure can no longer be defined as a situation in which the safety factor becomes less than unity; it must rather be defined with reference to excessive permanent displacements which impede the proper functioning of the structure. Although this definition seems rather simple and the methodology has been successfully applied to dam engineering (Newmark, 1965), its implementation in a code format is far from being an easy task. One of the difficulties is to define acceptable displacements of the structure in relation to the required performance. Another difficulty obviously lies in the uncertainty linked to the estimation of permanent displacements.

5.1 Code approach

As an example of code documentation Eurocode 8 states that "The stability against seismic bearing capacity failure taking into account load inclination and eccentricity arising from the inertia forces of the structure as well as the possible effects of the inertia forces in the supporting soil itself can be checked with the general expression and criteria provided in annex F. The rise of pore water pressure under cyclic loading should be considered either in the form of undrained strength or as pore pressure in effective stress analysis. For important structures, non linear soil behavior should be considered in determining possible permanent deformation during earthquakes."

More specifically, in most seismic codes the design engineer is required to check the following general inequality :

$$S_{\mathrm{d}} \leq R_{\mathrm{d}} \tag{1}$$

where S_{d} is the seismic design action and R_{d} the system design resistance. These two terms are explained below.

The design action represents the set of forces acting on the foundations. For the bearing capacity problem, they are composed of the normal force N_{sd}, shear force V_{sd}, overturning moment M_{sd} and soil inertia forces F developed in the soil. The actions N_{sd}, V_{sd}, and M_{sd} arise from the inertial soil-structure interaction. The inertia force, $F = \rho\, a$ (ρ mass density, a acceleration), arises from the site response analysis and kinematic interaction. The term design action is used to reflect that these forces must take into account the actual forces transmitted to the foundation i.e. including any behavior and over-strength factors used in inelastic design.

The design resistance represents the bearing capacity of the foundation; it is a function of the soil strength, soil-foundation interface strength and system geometry (for instance foundation width and length). Obviously, inequality (1) must include some safety factors. One way is to introduce partial factors, as in Eurocode 8. This is not the only possibility and some other codes, like the New Zealand one, choose the Load and Resistance Factored Method (LRFD) and factor the loads and resistance (Pender, 1999). The Eurocode approach is preferred because it gives more insight in the philosophy of safety; on the other hand it requires more experimental data and numerical analyses to calibrate the partial factors.

With the introduction of partial factors inequality (1) is modified as follows:

$$S_{\mathrm{d}}\left(\gamma_{\mathrm{F}}\,\text{actions}\right) \leq \frac{1}{\gamma_{\mathrm{Rd}}} R_{\mathrm{d}}\left(\frac{\text{strength parameters}}{\gamma_{\mathrm{m}}}, \text{geometry}\right) \tag{2}$$

where "actions" represent the design action and "strength" the material strength (soil cohesion and /or friction angle, soil-foundation friction coefficient).

- γ_F is the load factor applied to the design action: γ_F is larger than one for unfavorable actions and smaller than 1.0 for favorable ones.
- γ_m is the material factor used to reflect the variability and uncertainty in the determination of the soil strength. In Eurocode 8, the following values are used: 1.4 on the undrained shear strength and cohesion and 1.25 on the tangent of the soil friction angle or interface friction coefficient.
- γ_{Rd} is a model factor. It acts like the inverse of a strength reduction factor applied to the resistance in an LRFD code. This factor reflects the fact, that to evaluate the system resistance some approximations must be made: a theoretical framework must be developed to compute the resistance and like any model it involves simplifications, and assumptions which deviate from reality. It will be seen later on that the model factor is essential and can be used with benefit to differentiate a static problem from a seismic one.

5.2 Theoretical framework for the evaluation of the foundation bearing capacity

Since the devastating foundation failures reported after the Mexico earthquake (Auvinet and Mendoza, 1986) a wealth of theoretical and experimental studies have been carried out to develop bearing capacity formulae which include the effect of the soil inertia forces (Sarma and Iossifelis 1990, Budhu and Al Karni, 1993, Richards et al 1993, Zeng and Steedman, 1998). The theoretical studies mentioned above are based on limit analysis methods (Chen 1975, Salençon 1983); although they represent a significant improvement over the previous analyses, which neglected the soil inertia forces, they suffer from limitations that restrict their use (Pecker, 1994):

- the horizontal accelerations of the soil and of the structure are assumed to have the same magnitude;
- the results are derived from an assumed unique failure mechanism that does not allow for foundation uplift;
- the methods only consider upper bound solutions without any indication on how close they are to the exact solution.

At the same time numerous studies have been initiated in France and Europe with the objective of providing more general solutions (Pecker and Salençon 1991, Dormieux and Pecker 1994, Salençon and Pecker 1994 a-b, Paolucci and Pecker 1997, PREC8, 1996). The solutions were developed within the framework of the yield design theory (Salençon 1983, 1990): the loading parameters *N*, *V*, *M* and *F* are considered as independent loading parameters thereby allowing for any combination of actions to be analyzed; many different kinematic mechanisms are investigated and lower bound solutions are also derived to (i) obtain the best possible approximation to the bearing capacity, (ii) bracket the true value to obtain a quantitative measure of the goodness of the solution. It is interesting to note that the results have been later completed by additional lower bound solutions which confirm the merit of the upper bound solutions and help to narrow the gap between upper and lower bound solutions (Ukritchon et al, 1998). Finally the results, mainly based on the upper bound solutions are cast in the general format (Pecker, 1997):

$$\phi(N,V,M,F) \leq 0 \tag{3}$$

where $\phi(\ ) = 0$ (Figure 6) defines in the loading parameter space the equation of a surface defining the ultimate loads of the system.

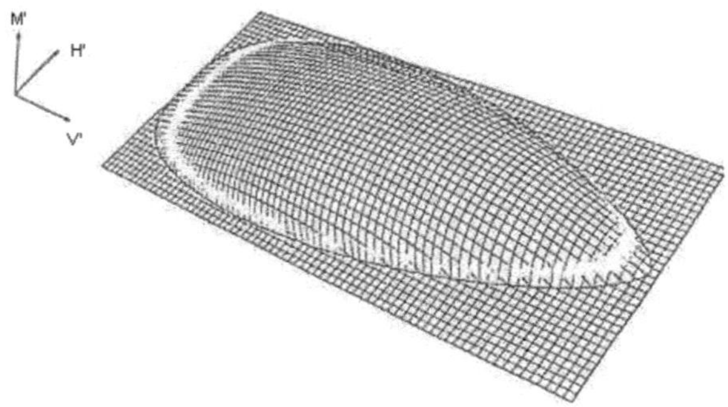

Figure 6 : Ultimate loads surface for cohesive soils

Inequality (3) expresses the fact that any combination of the loading parameters lying outside the surface corresponds to an unstable situation; any combination lying inside the bounding surface corresponds to a potentially stable situation. The word potentially is used to point out that no assurance can be given since the solutions were derived from upper bound solutions. Indications on the merit of the solutions are obtained by comparison with the lower bound solutions and the model factor of Eq.(2) is introduced to account for that uncertainty. The uncertainty is twofold: the solution is obtained from an upper bound approach and, although various kinematic mechanisms were investigated, their number remains necessarily limited when a comprehensive implementation of the upper bound theorem would require that all the conceivable mechanisms be investigated.

These results are put in a simple mathematical expression and implemented in the current version of Eurocode 8 (Annex F) and are applicable to cohesive and purely frictional materials. The equation, (Pecker, 1997), is:

$$\frac{\left(1-e\overline{F}\right)^{c_T}\left(\beta\overline{V}\right)^{c_T}}{\left(\overline{N}\right)^{a}\left[\left(1-m\overline{F}^{k}\right)^{k'}-\overline{N}\right]^{b}}+\frac{\left(1-f\overline{F}\right)^{c'_M}\left(\gamma\overline{M}\right)^{c_M}}{\left(\overline{N}\right)^{c}\left[\left(1-m\overline{F}^{k}\right)^{k'}-\overline{N}\right]^{d}}-1\leq 0 \qquad (4)$$

where : $\overline{N}=\frac{\gamma_{Rd}N_{Sd}}{N_{max}}$, $\overline{V}=\frac{\gamma_{Rd}V_{Sd}}{N_{max}}$, $\overline{M}=\frac{\gamma_{Rd}M_{Sd}}{BN_{max}}$

N_{max} is the ultimate bearing capacity of the foundation under a vertical centered load, N_{Sd}, V_{Sd}, and M_{Sd} are the design action effects at the foundation level, B the foundation width, and γ_{Rd} the model factor. The soil inertia forces are accounted for by the normalized parameter $\overline{F}$ equal to $\rho aB/c_u$ for cohesive soils and to $a/g\tan\phi$ for frictional soils. The other parameters entering equation (4) are numerical parameters derived by curve fitting to the "exact" bearing capacity, the values of which can be found in Pecker, 1997. Recent ongoing research (Chatzigogos *et al*, 2006) has shown that equation (4) is still approximately valid for a circular foundation provided the value of N_{max} for circular foundation is used.

5.3 Evaluation of permanent displacements

As noted previously and as recommended in Eurocode 8, in seismic situations, the permanent displacements should be evaluated. However such an evaluation is anything but an easy task. Probably the most rigorous approach would be to use a global model (finite element model) including both the soil and the structure. Obviously, the results depend on the non linear constitutive relationship used to model the soil behavior and are only meaningful if a realistic model is used. Owing to this constraint, to computer limitations, and to the required skill from the analyst in geotechnical engineering, structural engineering, soil-structure interaction and numerical analysis, such an approach is seldom used in everyday practice.

The alternative approach, once the seismic forces are known, is to rely on a Newmark type of approach (Newmark, 1965). The bounding surface defined by Eq.(3) is used as the surface defining the onset of permanent displacements. Sarma and Iossifelis (1990), Richards et al (1993) used the Newmark's approach assuming that the soil moves together with the foundation in a rigid body motion. The method has been further extended by Pecker and Salençon (1991) considering a deformable soil body corresponding to the assumed kinematic failure mechanism. Using the kinetic energy theorem, these authors computed the foundation angular velocity, and by integration over time, the foundation permanent rotation. When applied to actual case histories the method proved to be reliable (Pecker et al, 1995).

A potential use of the method can be found for the development of a code like approach. Computed permanent displacements develop when the resultant of the design action lies outside the bounding surface: the larger the distance to the bounding surface, the greater the displacements. This can be expressed mathematically by writing that for such situations:

$$S_{\mathrm{d}} = \lambda R_{\mathrm{d}} \qquad (5)$$

with $\lambda > 1$; $\lambda = 1$ corresponds to the onset of permanent displacements.

Comparing Eq.(5) to Eq.(2), it is readily apparent that allowing S_d to reach regions outside the bounding surface is equivalent to specifying a model factor γ_{Rd} smaller than 1.0. Therefore, γ_{Rd} can be used, in addition to reflecting the uncertainties in the model, to relax the constraint that at any time the resistance shall be larger than the action, recognizing the fundamental difference between a static problem and a seismic one in which forces vary in time.

This approach has been implemented in Eurocode 8 and the tentative values proposed in its Annex F are intended to allow for the development of small permanent displacements in potentially non dangerous materials (medium to dense sand, non sensitive clay). These values range from 1.0 (medium dense to dense sands, non sensitive clays) to 1.5 (loose saturated sands) with intermediate values of 1.15 for loose dry sands. If this phenomenon was disregarded, γ_{Rd} values would always be larger than 1.0 (in the range 1.2 to 1.5). In the case of non sensitive clays further justification for setting γ_{Rd} equal to 1.0 is the observation that shallow foundations in clay have generally been observed to perform well under seismic loading. A reason for this may be the enhanced undrained shear strength available under rapid loading (Romo 1995, and Ahmed-Zeki et al 1999).

5.4 Unresolved issues and further developments

One of the strong assumptions underlying the seismic bearing capacity checks is the independence between the computed design actions and soil yielding. Except for the sophisticated approaches involving the partition in near and far fields, the design actions are computed assuming quasi-linear foundation behavior. However it is recognized that partial yielding of the foundation may affect the forces.

Attempts have been made by Nova and Montrasio (1991) for monotonic static loading based on the concept of a macro-element modeling the soil and foundation; the constitutive law for the macro-element is rigid plastic strain hardening with non associated flow rule. That concept of macro-element expressed in global variables at the foundation level has been extensively used in mechanics but seldom applied to soil-structure interaction. Paolucci (1997) and Pedretti (1998) have extended the method to seismic loading. These last two studies definitively prove that yielding of the foundation cannot be ignored in the evaluation of the design action.

A more general formulation has been proposed recently by Cremer et al (2000). The developed macro-element taking advantage of the partition between near field and far field describes the cyclic behavior of the foundation, reproduces the material non-linearities under the foundation (yielding) as well as the geometrical non-linearities (uplift), and accounts for wave propagation in the soil. The strength criterion for the macro-element is represented by the bounding surface defined by the bearing capacity formula and a non associated flow rule with kinematic and isotropic hardening is used to compute the pre-failure displacements; the plastic model is coupled with an uplift model to integrate the influence of soil yielding on the uplift. Although still under development the model shows some promising capabilities and should represent a step forward in the evaluation of permanent seismic displacements of shallow foundations.

6 Stiffness and Capacity of Pile Foundations

As for shallow foundations, the ultimate capacity of the piles, or pile groups, has to be checked once the applied inertia loads acting at the pile cap are known. Two failure modes must be examined: bearing failure with a vertical force exceeding the available tip and shaft resistance and lateral failure when the available lateral resistance is mobilized along part the pile shaft. The latter failure mode is more likely to occur although bearing capacity failures of floating piles have been observed, for instance in Mexico City in the 1985 earthquake.

As well as capacity, the stiffness of the pile, or pile group, needs to be evaluated for the calculation of kinematic and inertial interaction. Kinematic effects are known not to have a great effect on pile head motions. The pile head lateral and rotational stiffness is highly nonlinear.

6.1 Single pile stiffness

Dealing first with the stiffness and considering initially a single vertical pile, we need the stiffness of the pile head to vertical, lateral and moment loading; in addition we need to be able to estimate the response of the pile shaft to soil movements which occur in kinematic interaction. The linear response is well documented. Two models are commonly used: the elastic continuum and the Winkler spring. There is an important difference between the vertical stiffness of a pile, particularly if the length of the pile is modest and it bears on a firm stratum, and the lateral stiffness. The axial stiffness of a vertical pile is larger than the lateral stiffness; it may involve soil-pile interaction over the full length of the pile shaft, whereas the lateral stiffness mobilizes a relatively short portion of the pile shaft.

Elastic continuum

The deflected shape of a pile which has horizontal lateral load applied at the ground surface extends a distance of several pile diameters into the soil profile. Beneath this there is negligible lateral displacement. The length over which the lateral displacement occurs is known as the active length and it is a function of pile diameter, the elastic modulus of the soil, and the ratio of pile modulus to soil modulus. Expressions for the active length for both static and dynamic lateral loading are given by Gazetas (1991). If the length of pile shaft is greater than this amount then the components of the pile head stiffness matrix are independent of the length of the pile shaft. Gazetas (1991) also gives expressions for the components of the pile head stiffness matrix for a variety of soil modulus distributions with depth; these are included in Eurocode 8. A very important finding about the dynamic lateral pile head stiffness is that the values are only slightly affected by loading frequency, thus a very good first approximation to the pile head stiffness components is given by the static stiffness of the pile. For dynamic loading there is, of course, damping. It has been found that for frequencies less than the natural period of the clay layer there is no radiation damping so the only damping is material damping in the soil adjacent to the pile. For frequencies higher than the natural frequency of the layer radiation damping, which increases with increasing frequency, is added to the material damping. Gazetas (1991) gives expression for the frequency effect on pile head stiffness and also expressions for the radiation damping. Thus, as stated above, simple equations are available for the elastic dynamic pile head stiffness coefficients which are needed for inertial interaction. The potential limitation that these are available only for homogeneous or simple variations of soil modulus with depth is not such a problem as the depth of soil involved in the interaction is limited by the active length of the pile shaft. Gazetas also provided simple expressions for the vertical stiffness and radiation damping for piles. However, the appropriateness of these may be limited by layering in the soil profile.

Winkler model

When dealing with layered soil profiles the elastic continuum model is not applicable, but, fortunately, the Winkler model is useful for this case. This model consists of a series of independent horizontal (lateral) or vertical (axial) springs distributed along the beam (pile) length. It has long been used in offshore engineering to compute the pile head deflection and settlement under static loading at the pile head (Matlock and Reese, 1960). Based on field experiments, the approach models the soil response at a particular depth in terms of a p-y curve for lateral loading or t-z curve for axial loading; p and y (t, z) denote the resultant lateral (axial) soil reaction per unit length of pile and the associated lateral (axial) pile deflection. Matlock and Reese give a method of estimating the spring stiffness based on laboratory test data on soil samples. Another approach is given in the French code using a tri-linear backbone curve for static design of foundations (Fascicule 62, 1992). The slopes of the three portions are related to the pressuremeter modulus (Menard's modulus). The two break points in the distributed force per unit length of the pile shaft occur at p_fD, and p_lD, where p_f and p_l are the creep and limiting pressures measured in a pressumeter test and D the pile diameter. In this approach a reduction is applied to the parameters near the ground surface because it is very unlikely that the soil near the ground surface contributes fully to the pile resistance. A further alternative, (Pender, 1993), uses an empirical approach to estimate the initial spring stiffness, evaluates the maximum lateral pressure from the shear strength of the soil, and has a hyperbolic transition between these two limits. For dynamic loading the soil-pile interaction is modeled not only with distributed springs but also with dashpots, which are frequency independent. Many different methods are available to determine the springs and dashpots in the linear, or quasi linear, case (Gazetas et al, 1992; Makris and Gazetas, 1992; Kavvadas and Gazetas, 1993).

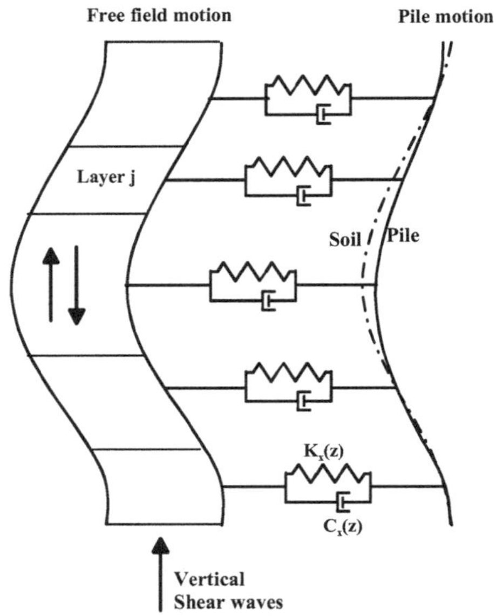

Figure 7 : Dynamic Winkler model for soil-pile interaction analyses

The Winkler model is also used for calculating kinematic interaction between a pile and a soil profile. The difference between Matlock and Reese's model and this dynamic model stems from the location of the applied excitation; in the dynamic model (Figure 7) the extremities of the springs and dashpots are connected to the free field where the soil response, computed independently, is imposed; in the static model those extremities are fixed. These calculations are presented by Kavvadas and Gazetas (1993) and Makris and Gazetas (1992). Tabesh and Poulos (1999) suggest that quite useful design information can be obtained for kinematic effects using an essentially static analysis of the pile shaft provided information about the free field displacement profile is available; they use a one dimensional site response analysis to obtain this.

Along the same lines as for the shallow foundation, the Winkler model can be extended to non linear situations. The springs and dashpots of Figure 7 model the far field domain. The near field domain is modeled with non linear, hysteretic, springs. The back bone curve for these springs would be the p-y curves, or any experimentally based curve for which the series of springs and sliders of Figure 5 constitute a piecewise linear approximation. Gapping is a nonlinear phenomenon and the Winkler model was used to calculate it. Dobry and Gazetas (1984) provide a simplified method for estimating nonlinear pile head stiffness and damping in layered soil deposits; they use a Winkler approach to estimate the pile head lateral stiffness and suggest a method for estimating the damping based on the deflected shape of the pile shaft. An alternative to calculations with a nonlinear Winkler spring model is equations given by Davies and Budhu (1986) and Budhu and Davies (1987) which give nonlinear relations between lateral load and pile head displacement and rotation. These relations were developed for static lateral loading of piles; the justification for using them in a dynamic context is finding that the pile head stiffness is not greatly affected by frequency.

Single pile capacity

The vertical capacity can be estimated using standard methods. For dynamic loads the capacity is likely to be greater than static because of enhanced soil strength mentioned above. Similarly there are standard methods for estimating lateral capacity. However for earthquake loading there are likely to be simultaneous vertical and lateral cyclic loads so the question of the capacity under this combined loading arises. As the lateral capacity is generated in the upper part of the pile shaft, over a length similar to the active length used for stiffness, one could allocate this length to generating lateral resistance – even if this is likely to be expensive in most cases. Another issue is the effect of the cyclic vertical capacity on the pile shaft. It is well known that severe cyclic degradation of the pile shaft capacity can occur, although this is unlikely to be serious as long as there is no reversal of shear stress along the pile shaft.

Pile group stiffness

Once again the need is to consider stiffness for inertial and kinematic interaction as well as group capacity. No special kinematic interaction calculation is necessary for pile groups as it has been found from numerical investigations that kinematic effects for pile groups are little different from those for individual piles. In evaluating the dynamic stiffness of a pile group the interactions between the piles are important, in just the same way as they are for the static stiffness of a pile group. Early investigations using boundary element calculations (Kaynia and Kausel, 1982) revealed frequency dependence rather more complex than that for a single pile. Fortunately, a quite remarkable simplification was discovered (Dobry and Gazetas, 1988) which enables the dynamic stiffness components of a pile group to be evaluated by means of simple formulae, the essence of which is representing the interaction between the piles in the group by the propagation of cylindrical waves. Thus the expressions for the interaction coefficients between the piles are given in terms of pile spacing, excitation frequency, and the velocity of waves through the soil between the piles. If one models a building on a pile group foundation as an equivalent single degree of freedom system using the Dobry and Gazetas equations to handle the pile group, then it is found that the stiffness of the system exhibits much less frequency dependence than the pile group alone, Pender (1993). If the response spectrum method is used to estimate the maximum building displacement it is found that the values obtained using simply the static stiffnesses are very similar to those obtained using the frequency dependent stiffness and damping terms for the pile group.

Pile group capacity

Under earthquake loading a pile group may be subjected to horizontal shear and moment loading in addition to changes in any static vertical load it will be required to carry. This suggests a surface similar to that in Figure 6 for a shallow foundation. Such a treatment has not been developed and would involve interactions between the vertical capacity of the pile group, shear capacity and moment capacity. In many applications the interaction between vertical and moment capacity is likely to be more critical as the moment resistance, or the greater part of it, comes from the axial capacity of the piles. Some preliminary work has been done on this for 2x2 and 3x3 pile groups by Pender (1993). Any moment capacity using the axial capacity of the pile shafts is subject to the potential degradation of the capacity mentioned above for single piles. This is not likely to be serious as long as there is no reversal of the shear stress on the pile shaft; whether or not reversal occurs is dependent on the ratio of the static vertical load carried by the group to the cyclic moment. Horizontal shear can often be equilibrated with lateral earth pressures at the sides of the raft rather than being transferred to the pile shafts.

7 Construction Detailing

Although the safety of a constructed facility does not rely only upon a blind application of seismic codes and standards which are used for its design and construction, those documents help significantly to minimize the most commonly encountered causes of deficiencies and failures. Because all phenomena described previously cannot be analyzed with the necessary mathematical rigor and are not often relevant to even sophisticated calculations, construction detailing must always be enforced in seismic design of foundations. This is one of the major merits of seismic codes.

Many of these detailing practices, which are found in the most recent codes, are little more than common sense and by no means, the points raised herein constitute an exhaustive list. However based on the author's experience, they represent the most common mistakes made in design by non-experienced designers :

- Foundations must not be located close to (or across) major active faults. Ground motions in the near field are far from being predictable and attempts to design buildings to accommodate such movements, especially the static co-seismic displacements associated with fault rupture, are almost hopeless.
- Liquefiable deposits and unstable slopes must always be treated before construction. Even if a piled foundation can survive the cyclic deformations of a liquefied deposit, the quasi static displacements imposed by the post-earthquake ground flow are an order of magnitude larger and cause distress of the foundation as evidenced in the Kobe earthquake.
- The foundation system under a building must be as homogenous as possible unless construction joints are provided in the structure. In particular, for individual footings, the situation where some of them rest on a man-made fill and some on in-situ soils must always be avoided. It is also highly desirable that the foundations respect the symmetries of the building.
- The choice of the foundation system must always account for possible secondary effects such as settlements in medium-dense or loose dry sands, the post-earthquake consolidation settlements of clay layers, the settlements induced by the post-earthquake dissipation of pore pressures in a non liquefiable sand deposit. Raft foundations or end bearing piles are to

be preferred whenever the anticipated magnitude of the settlements is high or when they can be highly variable across the building.

- Individual footings must always be linked with tie beams at the foundation level. These longitudinal beams must be designed to withstand the differential settlements between the footings.
- Piles must be reinforced along their whole length, even if calculations do not require reinforcement. Special care must be given to the connections with the raft or to soil layers interfaces when two layers in contact present marked differences in stiffness. For instance, the connection with the raft can be detailed to allow for a plastic ductile hinge as allowed in Eurocode 8.
- Inclined piles must preferably be avoided: they can be subjected to parasitic bending stresses due to soil densification following an earthquake, they induce large forces onto pile cap and, if their arrangement is not symmetric, permanent rotations may develop due to different stiffness of the pile group in each direction of loading.

References

Ahmed-Zeki, A. S., Pender, M J, and Fitch, N. R. (1999) "Strain rate effects on the undrained shear strength of Waitemata Residual Clay", Proc. 8th Australia New Zealand Conference on Geomechanics, Hobart, Vol. 2, pp. 791-796.

Applied Technology Council (1996). "ATC 40 : The seismic evaluation and retrofit of concrete buildings". 2 volumes. Redwood City CA.

Auvinet G, Mendoza M.J. (1986). "Comportamiento de diversos tipos de cimentacion en la zone lascustre de la Ciudad de Mexico durante el sismo del 19 de septiembre de 1985". Proceedings Symposium: Los Sismos de 1985; Casos de Mecanica de Suelos, Mexico.

Budhu, M and Davies, T G (1987) Nonlinear analysis of laterally loaded piles in cohesionless soils. Canadian Geotechnical Jnl., Vol. 24, pp. 289-296.

Chen W.F. (1975). "Limit analysis and soil plasticity". Elsevier Science Publishing Company.

Cremer C., Pecker A., Davenne L. (2000). "Elaboration of a SSI macro-element with uplift of shallow foundations". Implication of Recent Earthquake on Seismic Risk. Elnashai-Antoniou Eds., Imperial College Press, pp 127-138.

Cremer C., Pecker A., Davenne L. (2000). "Cyclic macro-element for soil structure interaction - Material and geometrical non linearities". Submitted for publication *Numerical Methods in Geomechanics.*

Davies, T G and Budhu, M (1986) Nonlinear analysis of laterally loaded piles in heavily overconsolidated clay, Geotechnique, Vol. 36 No. 4, pp. 527-538.

De Barros F.C., Luco E. (1990). "Discrete models for vertical vibrations of surface and embedded foundations". Earthquake Engineering and Structural Dynamics, vol. 19.

Dobry R., Gazetas G. (1988). "Simple methods for dynamic stiffness and damping of floating pile groups". Geotechnique, vol. 38, n° 4.

Dormieux L., Pecker A. (1995). "Seismic bearing capacity of a foundation on a cohesionless soil". Journal of Geotechnical Engineering, ASCE, Vol 121, pp 300-303.

Eurocode 8-Part 5. "Foundations, retaining structures and geotechnical aspects". Ref.No prEN 1998-5:1999

Fascicule 62 - Titre V du CCTG - Règles Techniques de Conception et de Calcul des Fondations des Ouvrages du Génie Civil, avril 1992.

Gazetas (1984). "Seismic response of end bearing piles". Soil Dynamics and Earthquake Engineering, Vol 3, N°2, pp 82-93.

Gazetas, G and Dobry, R (1984) Horizontal response of piles in layered soils, Jnl. Geotech. Eng., Proc. ASCE, Vol. 110 No. 1, pp. 20-40.

Gazetas G. (1990). "Foundation vibration". Foundation Engineering Handbook, Chap. 15, 2nd Edition - Hsai-Yan Fang Eds.

Gazetas G., Fan, K., Tazoh T., Shimizu K., Kavvadas M., Makris N., (1992). "Seismic Pile-Group-Structure Interaction". Piles under dynamic Loads, ASCE, S. Prakash Ed., pp. 56-93.

Gazetas G., Mylonakis G. (1998). "Seismic soil structure interaction : new evidence and emerging issues". Geotechnical Earthquake Engineering and Soil Dynamics, ASCE, II, pp 1119-1174.

Halling, M W, Womack, K C, Muhamad, I and Rollins, K M (2000) Vibrational testing of a full-scale pile group in soft clay. Proc. 12th World Conference on Earthquake Engineering, Auckland. Paper 1745.

Idriss I.M., Seed H.B. (1968). "Seismic response of horizontal soil layers". Journal of Soil Mechanics and Foundation Division, ASCE, vol. 96, N° SM4.

Kavvadas M., Gazetas G., (1993). "Kinematic Seismic Response and Bending of Free-Head Piles in Layered Soil". Geotechnique, Vol. 43, N° 2, pp 207-222.

Kausel E., Roesset J.M. (1974). "Soil Structure Interaction for Nuclear Containment Structures". Proc. ASCE, Power Division Specialty Conference, Boulder, Colorado.

Lysmer J. (1978). "Analytical procedures in soil dynamics". Earthquake Engineering and Soil Dynamics Vol. III, Pasadena - Ca, pp. 1267-1316.

Makris N., Gazetas G., (1992). "Dynamic Pile-Soil-Pile Interaction. Part II : Lateral and Seismic Response". Earthquake Engineering and Structural Dynamics, Vol. 21, pp 145-162.

Masing G. (1926). "Eigensprannung und Verfestigung beim Messing" Proceedings International Congress of Applied Mechanics.

Matlock H., Reese L.C. (1960). "Generalized Solutions for Laterally Loaded Piles". Journal of Soil Mechanics and Foundation Division, ASCE, Vol 86, SM5, pp. 63-91.

Newmark N. (1965). "Effects of earthquakes on dams and embankments". Geotechnique, Vol. XV(2), pp 139-160.

Nova R., Montrasio L. (1991). "Settlements of shallow foundations on sand". Geotechnique 41, n° 2,pp. 243-256.

Paolucci R. (1997). "Simplified evaluation of earthquake induced permanent displacement of shallow foundations". Journal of Earthquake Engineering, Vol 1, n°3, pp 563-579.

Paolucci R., Pecker A. (1997). "Seismic bearing capacity of shallow strip foundations on dry soils". Soils and Foundation, Vol 37, n°3, pp 95-105.

Pecker A., Salençon J. (1991). "Seismic bearing capacity of shallow strip foundations on clay soils". CENAPRED, Proceedings of the International Workshop on Seismology and Earthquake Engineering, Mexico, pp 287-304.

Pecker A. (1995). "Seismic design of shallow foundations". State of the Art: 10th European Conference on Earthquake Engineering, Duma Ed., Balkema, pp 1001-1010.

Pecker A., Salençon J., Auvinet G., Romo M.P., Verzurra L. (1995). "Seismic bearing capacity of foundations on soft soils". Final Report to European Commission - Contract CI1 - CT92-0069.

Pecker A. (1997). "Analytical formulae for the seismic bearing capacity of shallow strip foundation". Seismic Behavior of Ground and Geotechnical Structures, Seco e Pinto Ed., Balkema, pp 261-268.

Pecker A. (1998). "Rion Antirion Bridge - Lumped parameter model for seismic soil structure interaction analyses - Principles and validation". Geodynamique et Structure Report FIN-P-CLC-MG-FOU-X-GDS00060 - Prepared for Gefyra Kinopraxia.

Pedretti S. (1998). "Nonlinear seismic soil-foundation interaction: analysis and modelling method". PhD Thesis Dpt Ing Structurale, Politecnico di Milano.

Pender M. (1993). "Aseismic Pile Foundation design Analysis". Bulletin of the New Zealand National Society for Earthquake Engineering, Vol. 26, No 1, pp 49-160.

Pender, M J (1994) Components of the stiffness of pile raft foundations. Proc. XIII International Conference on Soil Mechanics and Foundation Engineering, New Delhi. Vol. 2, pp. 923-928.

Pender, M J (1994) "Earthquake response of structures on pile group foundations", Proc. 1st ROC - NZ Workshop on Earthquake Engineering, Taiwan, May, Huei-Tsyr Chen editor, Taiwanese National Centre for Research on Earthquake Engineering, pp 32 - 50.

Pender, M.J. (1995). "Earthquake Resistant design of Foundations". Keynote address Pacific Conference on Earthquake Engineering, PCEE95, Melbourne.

Pender M.J., Pranjoto S., (1996). "Gapping effects during cyclic lateral loading of piles in clay", Proceedings 11th World Conference on Earthquake Engineering - Acapulco.

Pender M.J. (1999). "Geotechnical Earthquake Engineering design practice in New-Zealand". Proceedings of the 2nd International Conference on Earthquake Geotechnical Engineering. Sêco e Pinto Ed., Balkema.

PREC 8 (1996). (Prenormative Research in Support of Eurocode 8) "Seismic behavior and design of foundation and retaining structures". Facioli-Paolucci Eds., Report n°2.

Richards R., Elms D.G., Budhu M. (1993). "Seismic bearing capacity and settlements of shallow foundations". Journal of Geotechnical Engineering, ASCE, Vol 119 n° 7, pp 662-674.

Romo, M. (1995) “Clay behaviour ground response and soil-structure interaction studies in Mexico City”, Proc. 3rd. Conf. on Recent Advances in Geotech. Earthq. Engng. and Soil Dynamics, Vol. 2, pp. 1039-1051.

Salençon J. (1983). "Calcul à la rupture et analyse limite" Presses de l'Ecole Nationale des Ponts et Chaussées, Paris.

Salençon J. (1990). "An introduction to the yield design theory and its application to soil mechanics". European Journal of Mechanics A/Solids, Vol 9(5), pp 477-500.

Salençon J., Pecker A. (1994a). "Ultimate bearing capacity of shallow foundations under inclined and eccentric loads. Part 1 : Purely cohesive soil". European Journal of Mechanics A/Solids, 14, n° 3, pp. 349-375.

Salençon J., Pecker A. (1994b). "Ultimate bearing capacity of shallow foundations under inclined and eccentric loads. Part II : Purely cohesive soil without tensile strength". European Journal of Mechanics A/Solids, 14, n° 3, pp. 377-396.

Sarma S.K., Iossifelis I.S. (1990). "Seismic bearing capacity factors of shallow strip footings". Geotechnique, Vol .40, pp 265-273.

Tabesh, A and Poulos, H G (1999) Kinematic versus static interaction of pile and soil, Proc. 8th Australia New Zealand Conference on Geomechanics, Hobart, Vol. 1, pp. 445-450.

Ukritchon B., Whittle A.J., Sloan S.W. (1998). "Undrained limit analysis for combined loading of strip footings on clay". Journal of Geotechnical and Geoenvironmental Engineering, March, pp 265-275.

Wolf J.P. (1994). "Foundation vibration analysis using simple physical models". Prentice Hall Inc.

Yasuda, S and Berrill, J B (2000) Observations of the earthquake response of foundations in soil profiles containing saturated sands, GeoEng2000.

Zeng X., Steedman R.S. (1998). "Bearing capcity failures of shallow foundations in earthquakes". Geotechnique, 48, n° 2, pp. 235-256.

Nonlinear Dynamic Analysis of Structures Subjected to Seismic Action

Rui Pinho

Department of Structural Mechanics, University of Pavia, Pavia, Italy

Abstract. It is widely recognized that nonlinear time-history analysis constitutes the most accurate way for simulating response of structures subjected to strong levels of seismic excitation. This analytical method is based on sound underlying principles and features the capability of reproducing the intrinsic inelastic dynamic behaviour of structures. Nonetheless, comparisons with experimental results from large-scale testing of structures are still needed, in order to ensure adequate levels of confidence in this numerical methodology. The fibre modelling approach employed in the current endeavour inherently accounts for geometric nonlinearities and material inelasticity, without a need for calibration of plastic hinges mechanisms, typical in concentrated plasticity models. The resulting combination of analysis accuracy and modelling simplicity, allows thus to overcome the perhaps not fully justifiable sense of complexity associated to nonlinear dynamic analysis. The fibre-based modelling approach is employed in the framework of a finite element program for seismic response analysis of framed structures, freely downloaded from the Internet. The reliability and the accuracy of the program are demonstrated by numerically reproducing pseudo-dynamic tests on full-scale structures. Modelling assumptions are discussed, together with their implications on numerical results of the nonlinear time-history analyses, which were found in good agreement with experimental results.

1 Introduction

Old generation of design codes based on equivalent elastic force approaches proved to be ineffective in preventing earthquake destructive consequences. After recent major earthquakes (e.g. Northridge 1994, Kobe 1995, Kocaeli 1999), the necessity for using ever more accurate methods, which explicitly account for geometrical nonlinearities and material inelasticity, for evaluating seismic demand on structures, became evident. Within this framework, two analysis tools are currently offered with different levels of complexity and of required computational effort; nonlinear static analysis (pushover) and nonlinear dynamic analysis (time-history). Even if the latter is commonly considered to be complex and not yet mature enough for widespread professional use, it constitutes the most powerful and accurate tool for seismic assessment. In the latest generation of seismic regulations, dynamic analysis of three dimensional structural models is indeed recommended for the assessment of existing critical structures in zones of high seismic risk, as well in the planning and design of appropriate retrofitting strategies.

Given its conspicuous increased computational effort and analytical complexity, the common use of nonlinear finite element (FE) methods in design applications requires first a clear demonstration of its accuracy and reliability. Within this scope, the current work tries to establish the ability of a fibre-modelling approach in predicting the seismic response of reinforced concrete structures, through the reproduction of pseudo-dynamic tests carried out at the Joint Research Centre of Ispra on a continuous span bridge (Pinto *et al.*, 1996) and a four-storey frame (Pinto *et al.*, 2002). The employed numerical algorithm allows for automatic accounting of both local (beam-column effect) and global (large displacements/rotations effects) sources of geometric nonlinearity, together with proper modelling of material cyclic inelasticity.

Further, the nonlinear analysis software package employed in this endeavour is freely downloadable from the Internet, thus giving readers the opportunity to try hands-on the proposed numerical scheme in a graphical-interfaced software package adequate for general use, and in this way overcome, at least partially, the sense of complexity associated to nonlinear dynamic analysis.

2 Modelling Approaches for Inelastic Analysis

Nonlinear time-history analyses are a very powerful tool, provided they are supported by proper approximations and modelling. The analysis is inherently complex and may be very time-consuming, depending on the choice of the integration time-step, of the integration scheme, of the nonlinear incremental iterative algorithm strategy, and of the size of the mesh: an optimum balance among all these features will cater for accurate solutions with relatively reduced computational effort.

As described by Spacone (2001), a suitable classification of the different modelling strategies available may be based on the objective of the numerical study.

1. In Global Models (or Lumped Parameters Models), the nonlinear response of a structure is represented at selected degrees of freedom.
2. In Discrete FE Models (also called Member Models, or Structural Elements Models, or Frame Models) the structure is characterised as an assembly of interconnected frame elements with either lumped or distributed nonlinearities.
3. Microscopic Finite Element Models use the FE general method of structural analysis, in which the solution of a problem in continuum mechanics is approximated by the analysis of an assemblage of two or three-dimensional FEs which are interconnected at a finite number of nodal points and represent the solution domain of the problem.

The level of refinement of the model depends on the required accuracy and on the available computational resources. While refined FE models might be suitable for the detailed study of small parts of the structure, such as beam-column joints, frame models are currently the only economical solution for the nonlinear seismic analysis of structures with several hundred members. In other words, member FE models are the best compromise between simplicity and accuracy, as they represent the simplest class of models that nonetheless manage to provide a reasonable insight into both the seismic response of members and of the structure as a whole.

Assumptions and simplifications on the model with respect to the real structure are necessary, but need careful consideration because of their influence on results, which must be critically analysed accordingly. In the particular case of bridges, for instance, the structural subsystems that may be potentially hit by intense seismic action are the deck, the bearing structure and the foundation system. Due to the cost and technical difficulties in its repairing, foundations are usually protected from damage, whilst for reasons of life safety, the deck is kept elastic (though cracking is inevitably allowed for). Indeed, the most common trend in earthquake-resistant design of bridges assigns therefore to the bearing structure, and by means of inelastic deformation mechanisms, a key role in dissipating the energy introduced by the earthquake loads, for which reason these are normally the elements requiring the most accurate modelling.

2.1 Representation of Inelasticity

Two different modelling philosophies are commonly employed in the analytical reproduction of the inelastic response of structures subjected to seismic action; the 'concentrated plasticity' and the 'distributed inelasticity' modelling approaches.

As stated by Spacone (2001), due to the typical concentration of inelasticity of RC frames at the extremities of it structural elements, "an early approach to modelling this behaviour was by means of nonlinear springs at the member ends (Clough and Johnston, 1966, Giberson, 1967, and Takizawa, 1976). Among the lumped plasticity constitutive models proposed, some include stiffness degradation in flexure and shear (Clough and Benuska, 1967, Takeda *et al.*, 1970, and Brancaleoni *et al.*, 1983), 'pinching' under load reversal (Banon *et al.*, 1981, Brancaleoni *et al.*, 1983), and fixed end rotations at the beam-column joint interface to simulate the effect of bar pull-out (Otani, 1974, Filippou and Issa, 1988)". Such concentrated plasticity approach should be used with care, since accuracy of the analysis may be compromised whenever users are not highly experienced in the calibration of the available response curves, needed to characterise the lumped plasticity elements. The limitations of lumped models are discussed in several studies, such as Charney and Bertero (1982) and Bertero *et al.* (1984), amongst others.

The 'distributed inelasticity' modelling describes more accurately the continuous structural characteristics of reinforced concrete members, requiring simply geometrical and material characteristics as input data. The constitutive behaviour of the cross-section can be either formulated according to the classical plasticity theory in terms of stress and strain resultants, or explicitly derived by discretising the cross section into fibres. The latter approach, known as "fibre modelling", represents the spread of material inelasticity both along the member length and across the section area, thus allowing an accurate estimation of the structural damage distribution even in the highly inelastic range.

Quoting Spacone (2001) again, "the first elements with distributed nonlinearity were formulated with the classical stiffness method using cubic hermitian polynomials to approximate the deformations along the element (Hellesland and Scordelis, 1981, Mari and Scordelis, 1984). Menegotto and Pinto (1973) interpolated both section deformations and section flexibilities and accounted for the axial force-bending moment interaction. Shear effects were first included in the model proposed by Bazant and Bhat (1977)." More recently, alternative flexibility-based formulations have been developed by Mahasuverachai and Powell (1982), Kaba and Mahin (1984), Zeris and Mahin (1988, 1991), however these posed difficulties with regards to their implementation in FE programs. To overcome such complications, Ciampi and Carlesimo (1986) proposed a consistent flexibility-based method for formulating frame member models, later applied by Spacone (1994) to the formulation of a fibre beam-column element. A detailed discussion on the differences between stiffness-based and flexibility-based approaches may be found in Papaioannou *et al.* (2005), for instance.

For the purpose of the current work, a classical stiffness-based formulation, as developed by Izzuddin (2001), has been adopted.

3 Fibre Modelling Approach

In fibre modelling, the sectional stress-strain state of the elements is obtained through the integration of the nonlinear uniaxial stress-strain response of the individual fibres in which the section is subdivided, distinguishing steel, confined and unconfined concrete, as illustrated in Figure 1. The adopted stiffness-based element cubic formulation then allows both the representation of the spread of inelasticity along the member length as well as the implicit incorporation of interaction between axial force and transverse deformation of the element. The use of a sufficient number of elements per structural member permits the reproduction of plastic hinge (in their full length), typical of members subjected to high levels of material inelasticity. The spread of inelasticity across the section and along the member length is thus achieved without requiring expertise calibration of any lumped plasticity element.

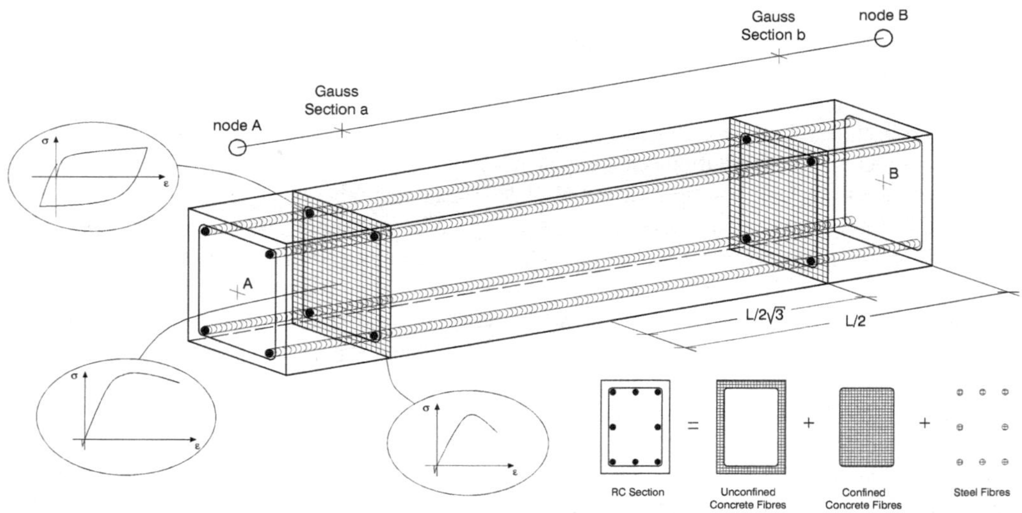

Figure 1. Discretization of a typical reinforced concrete cross-section

Structural members are represented by means of frame elements, with finite length and assigned cross-sections. Structural and non-structural inertia mass may also be introduced, in either lumped or distributed fashion, whilst joint/link elements, defined as spring-type elements joining coincident locations, can be used to model discontinuous connections. By means of such element types, a number of different element classes (columns, beams, walls, beam-column joints, etc.), non-structural components (energy dissipating devices, inertia masses, etc.) and different boundary conditions (flexible foundations, seismic isolation or structural gapping and pounding) can be represented.

The fibre-discretization renders possible a realistic modelling of the different materials, and their distribution, that make up a given member cross-section. The employable material models may feature different levels of accuracy/complexity in their definition; the bilinear, the Menegotto-Pinto (1973) and the Monti-Nuti (1992) models are among the most used models for steel, whilst concrete may be characterized by tri-linear, nonlinear with constant or variable confinement constitutive laws (see Scott *et al.*, 1982, Mander *et al.*, 1987). Many other material constitutive laws are available in the literature.

4 Assembling a FE Model for Dynamic Analyses for Reinforced Concrete Structures

Some recommendations for nonlinear analyses of reinforced concrete models can be summarised as follows:

i. At the onset of the development of a FE model, a sensitivity study will allow the attainment of reliable models: a too fine mesh may cause numerical instabilities, whilst, on the other hand, if the mesh is exaggeratedly coarse the analysis will not be sufficiently accurate. The meshing of the structure can be optimally carried out by refining critical structural locations, such as the zones where high inelasticity is expected or where abrupt changes in the stiffness of joined elements are present, such as plastic hinges locations, elements connections, and structural boundaries.

ii. The modelling of each structural element should be based on its expected behaviour: some examples are the linear behaviour of the deck or the modelling of the plastic hinge length to account for the flexibility of foundations. In this case, the use of inelastic fibre elements allows the explicit consideration of the spread of inelasticity.

iii. Model accuracy can be improved by using realistic materials property values and by properly defining boundary conditions. Considering soil-structure interaction allows a more realistic prediction of the seismic response of the model. In its simplest form, this may be implemented through boundary springs to which reasonable stiffness values, obtained from site investigation, are assigned.

iv. The use of preliminary eigenvalue analysis will assist in the verification of the correct assemblage of the model, in terms of stiffness allocation to structural members and mass distribution. In addition, the frequency characterisation of the structure is also commonly employed for the calibration of viscous damping, if the latter is deemed necessary, and will also provide some first insight into the expected response of the structure to a particular input motion.

4.1 Dynamic Analysis Features

As stated earlier, dynamic analysis is used to predict the nonlinear inelastic response of a structure subjected to earthquake loading: the seismic action may be introduced by means of acceleration loading curves at the supports, which may also be different at each support so as to represent asynchronous ground excitation. Mass and damping elements must be defined.

Dynamic analysis involves the direct integration of the equations of motion, which may be accomplished using the numerically dissipative-integration algorithm (Hilber *et al.*, 1977) or, as a special case of the latter, the well-known Newmark scheme (Newmark, 1959). The nonlinearity of the analysis scheme calls for the use of an incremental iterative solution procedure: this means that loads are applied in predefined increments, equilibrated through an iterative scheme, whereby the internal forces corresponding to a displacement increment are computed until either convergence is achieved or the maximum number of iterations is reached. At the completion of each incremental solution, before proceeding to the next load increment, the stiffness matrix of the model is updated to reflect nonlinear changes in structural stiffness. The solution algorithm may feature a hybrid incremental algorithm, obtained from a combination of the Newton-Raphson and the modified Newton-Raphson procedures, whereby the stiffness matrix is updated only in the first few iterations of a load step, thus obtaining an acceptable compromise between velocity in achieving convergence and required computational effort. The reader is referred, for instance, to the work of Cook *et al.* (1989) and Crisfield (1997) for further discussion on these topics.

In nonlinear analysis, automatic time-step adjustment controls load step sizes for optimum accuracy and efficiency: if the convergence is achieved easily, automatic time stepping will increase the load increment up to a selected maximum load step size, whilst if convergence is hard to achieve, automatic time stepping will bisect the load increment until a selected minimum load step size.

Different convergence check schemes, which may make use of three distinct criteria (displacement/rotation, force/moment, energy based), can be employed to check the convergence of a solution at the end of each iteration. The displacement/rotation criterion provides a direct local control over the precision obtained in the solution of the problem, usually ensuring overall accuracy. The force/moment criteria are suggested if a displacement convergence check is not sufficient for the internal forces of the elements to be adequately balanced. The maximum accuracy and solution control is obtained by combining the displacement and force convergence check criteria, while the maximum analysis stability is obtained if convergence is achieved for one of the two criteria checked, at a price of lower analytical precision. Tolerances in convergence criteria should be carefully defined.

5 Case Study 1: Multi-Span Continuous Deck Motorway Bridge

In the framework of an integrated European programme of pre-normative research in support of Eurocode 8 (CEN, 2002), six bridge prototypes, representative of typical multi-span continuous deck motorway bridges, have been designed (Pinto *et al.*, 1996) with different procedures for a PGA of *0.35g*, in medium soil conditions (soil type B), applying the EC8 provisions. Corresponding large-scale (1:2.5) bridge models have then been constructed and tested in pseudo-dynamic (PsD) fashion at the Joint Research Centre at Ispra (Italy).

A PsD test, despite being carried out quasi-statically, employs on-line computer calculations and control together with experimental measurement of the properties of the real structure, to provide a realistic simulation of its dynamic response. Inertial and viscous damping forces are modelled analytically, and an earthquake ground acceleration history is given as input data to the computer running the pseudo-dynamic algorithm. The horizontal displacements of the controlled degrees of freedom are calculated and then applied to the test structure by servo-controlled hydraulic actuators fixed to the reaction wall. The PsD testing of the bridge was performed using the sub-structuring technique, in which the piers were physically tested and the deck was numerically simulated on-line. Further details and references can be found in Pinto *et al.* (1996), Pinho (2000), Sullivan *et al.* (2004), amongst others.

5.1 Description of the Model

The tested bridge model labelled as B213C consists of three piers *5.6*, *2.8* and *8.4 m* high and a continuous deck with four identical *20 m* spans. For what concerns the boundary conditions, the deck is considered to end at the abutments with shear-keys, with the extremities free to rotate, as shown in Figure 2; the deck-piers connections are assumed to be hinged (no transmission of moments), transmitting lateral forces due to the engaging of the sub and superstructure in the transversal direction by means of the aforementioned shear keys.

The piers have rectangular hollow section with *160 mm* wall thickness (Figure 2). The minimum diameter of the longitudinal rebars and of the stirrups is 8 mm and 6 mm, respectively. The reinforcement layout of the pier models are shown in Figure 3. The mechanical characteristics of materials (B500 Tempcore steel with $E = 206\ GPa$ for longitudinal rebars and C25/30 concrete) and the mechanical characteristics of the pier cross-sections are shown in Table 1 and Table 2.

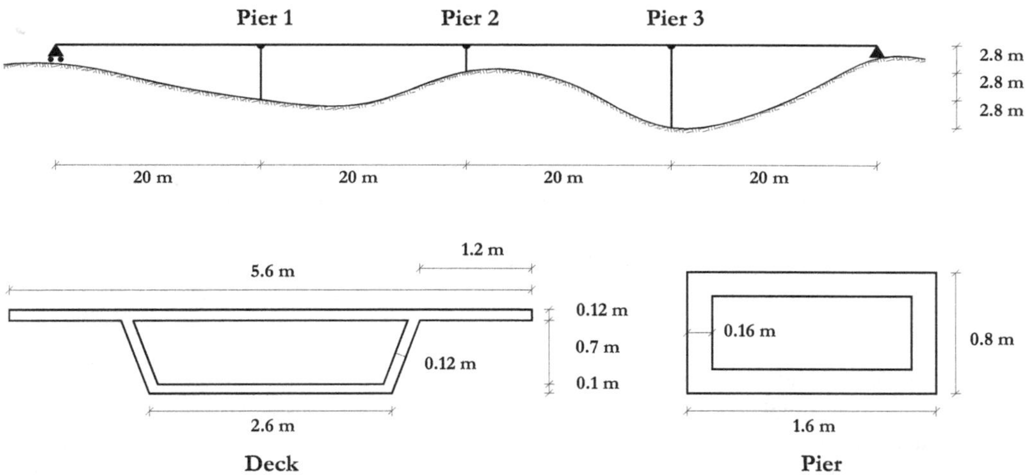

Figure 2. Bridge configuration and member cross sections

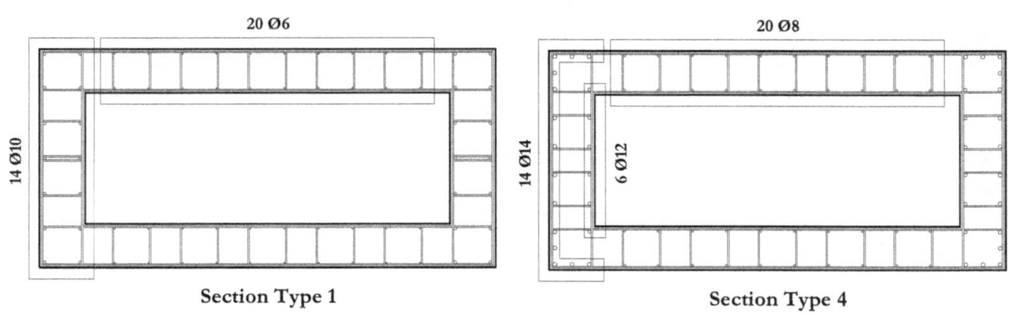

Figure 3. Reinforcement layout

Table 1. Steel mechanical properties (Guedes, 1997)

Diameter (mm)	Yield Strength (MPa)	Ultimate Strength (MPa)	Yield Strain (%)	Ultimate Strain (%)	Hardening
6	363.7	430.4	0.177	15.10	0.0022
8	503.4	563.0	0.244	12.30	0.0024
10	489.3	572.3	0.238	14.50	0.0028
12	558.2	646.8	0.271	12.80	0.0034
14	477.2	577.7	0.232	13.00	0.0038

Table 2. Summary of the pier cross section characteristics of the bridge (Guedes, 1997)

Pier	Section Type	Height (m)	Longitudinal steel (%)	Cubic concrete strength (MPa) (Compressive/tensile)
Pier 1	4	14	1.15	37.0 / 3.1
Pier 2	1	7	0.50	41.2 / 3.1
Pier 3	4	21	1.15	50.5 / 3.1

Table 3. Deck cross section geometrical and mechanical characteristics (Guedes 1997)

EA (kN)	EI_2 (kNm^2)	EI_3 (kNm^2)	GJ (kNm^2)
2.7837E+07	1.3544E+07	5.6517E+07	2.8017E+07

The deck is a hollow-core pre-stressed concrete girder *5.6 m* wide, as depicted in Figure 2. In the PsD test, it was simulated numerically with *32* linear elastic Timoshenko eccentric beam elements, whose mechanical characteristics are presented in Table 3, where A is the cross-section area, $I2$ and $I3$ are the two moments of inertia with respect to the local principal axes, J is the torsional constant and E is the Young Modulus of *25 GPa*. The inertia characteristics of the deck are based on a specific weight of $25\ kN/m^3$. As the deck is assumed to behave elastically, the substructured part included a Rayleigh damping matrix, featuring a damping ratio $\xi = 0.016$ associated to the two lower transversal natural frequencies of the complete bridge.

At the top of each pier, an axial force $N = 1700\ kN$ was applied by means of actuators, so as to simulate the vertical load that is transmitted from the deck. The input ground motion was represented by an adequately scaled accelerogram with duration of 4 seconds and a nominal peak acceleration of *0.875 g*, as shown in Figure 4. Two pseudo-dynamic tests were performed on the structure: one with the input motion corresponding to the design earthquake and another defined on the basis of the estimated ultimate capacity of the bridges, and thus equal to *1.2* times the design earthquake.

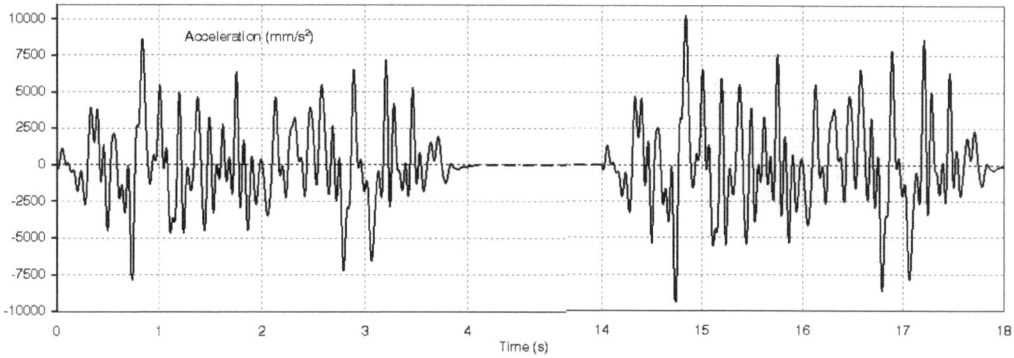

Figure 4. Input ground motion, design earthquake (Guedes, 1997)

5.2 Modelling of the Structure in the FE Program

The program employed in the current work, SeismoStruct (SeismoSoft, 2005), is a fibre–modelling FE package for seismic analysis of framed structures freely downloadable from the Internet. The program is capable of predicting the large displacement behaviour and the collapse load of any framed type of structural configuration under static or dynamic loading, accounting for geometric nonlinearities and material inelasticity.

As the structural details of a real structure are complex, simplification is needed in developing a FE bridge model, in order to produce predictions accurate enough with relatively reduced computational effort: a proper balance is required to avoid numerical instability and, on the other hand, to obtain results with a sufficient level of accuracy. Among the most common simplifications/assumptions requiring thoughtful consideration: (i) the structural mass is generally concentrated at the top of the pier, representing one single translational DOF, which however might not be accurate enough when the transversal size of the deck section is large with respect to the pier height, thus requiring an additional rotational DOF of the deck, (ii) the top concentrated mass assumption may be no longer acceptable when piers are massive with respect to the deck, (iii) the soil-structure interaction can be neglected or modelled with different levels of complexity, (iv) the influence of the shear deformation needs adequate analytical characterisation on squat members, for which shear collapse modes and flexure-shear interaction are relevant, (v) the penetration of plasticization at the base of piers may be modelled by extending the actual pier length, (vi) the influence of the spatial variability and/or loss of coherence of the ground motion may be represented by means of asynchronous input definition, (vii) connections among foundations, piers and deck can be modelled with different levels of complexity and detail.

In what follows, a description of geometry and discretisation of the model, its element connections, boundary conditions and loading state is given Adopted nonlinear analysis procedures and convergence criteria are also explained in some detail.

Modelling the bridge piers. As discussed previously, the piers are the elements in which inelastic deformation will be concentrated, therefore, a good level of accuracy in the characterization of materials and in the discretization of the mesh should be ensured. The piers have thus been modelled through a 3D inelastic beam-column element capable of capturing geometric and material nonlinearities. The number of fibres used in section equilibrium computations was set to 400; the selection of such number guarantees an adequate reproduction of the stress-strain distribution across the element cross-section, considering the shape and material characteristics of the latter, and the degree of material inelasticity that it is likely to reach.

The pier cross-section has thus been defined through a RC rectangular hollow section of *0.8 m* x *1.6 m*, with a wall width of *0.16 m*, a concrete cover of *8 mm*, and a steel layout reproducing the test specimen of Figure 3. The specimen sections contain steel rebars with different mechanical properties, as illustrated in Table 1, however, given the possibility of specifying only one steel material per section in the computer code used, an equivalent steel has been defined for each section (see Table 4), weighting its properties, i.e. yielding strength f_y and strain hardening parameters, proportionally to the distance from the sectional centre of gravity and to the area of each rebar.

Table 4. Equivalent steel properties per section

Section	Yield strength	Hardening
section 1	468 MPa	0.0027
section 4	496 MPa	0.0036

The stress-strain behaviour of the employed reinforcing steel (see Figure 5, left) has been described by the nonlinear model of Menegotto and Pinto (1973), as modified by Filippou *et al.* (1983) to include isotropic strain hardening. This is an accurate and convenient model, due to its computational efficiency and its very good agreement with experimental results. It utilises a damage modulus to represent more accurately the unloading stiffness, and has been modified and improved by Fragiadakis *et al.* (2006) to attain better stability and accuracy. The concrete has been represented through a nonlinear constant confinement concrete model (Figure 5, right), as a good compromise between simplicity and accuracy: it is an uniaxial nonlinear model following the constitutive relationship proposed by Mander *et al.* (1988), later modified by Martinez-Rueda

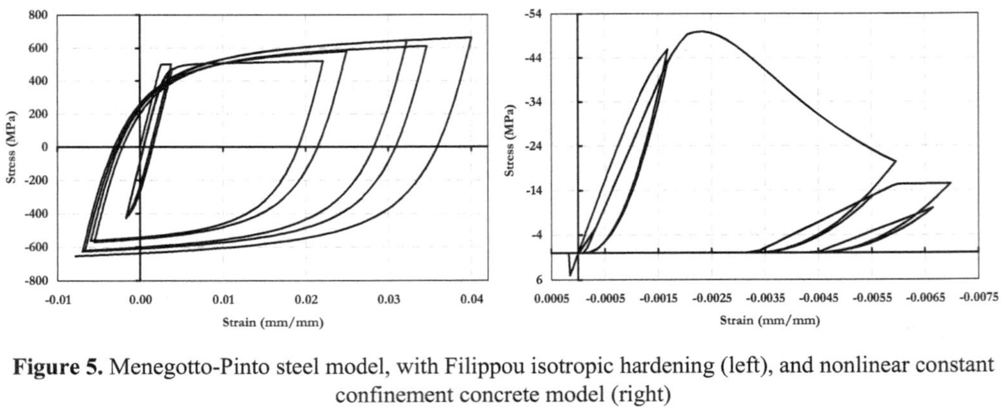

Figure 5. Menegotto-Pinto steel model, with Filippou isotropic hardening (left), and nonlinear constant confinement concrete model (right)

Table 5. Parameters for the Menegotto-Pinto steel model, with Filippou isotropic hardening

Parameter	**Sec 1**	**Sec 4**
Modulus of elasticity (MPa)	203000	203000
Yield strength (MPa)	468	496
Strain hardening parameter	0.0027	0.0036
Transition curve initial shape parameter (default value)	20	20
1st transition curve shape coefficient (default value)	18.5	18.5
2nd transition curve shape coefficient (default value)	0.15	0.15
1st isotropic hardening coefficient (default value)	0.025	0.025
2nd isotropic hardening coefficient (default value)	2	2

Table 6. Parameters for the nonlinear constant confinement concrete model

Parameter	Pier 1	Pier 2	Pier 3
Cylinder compressive strength (MPa)	31.5	35.0	42.9
Tensile strength (MPa)	3.1	3.1	3.1
Strain at unconfined peak stress (m/m)	0.002	0.002	0.002
Constant confinement factor	1.2	1.2	1.2

and Elnashai (1997) for reasons of numerical stability under large deformations. The constant confinement factor is defined as the ratio between the confined and unconfined compressive stress of the concrete. The model calibrating parameters, fully describing the mechanical properties of steel and concrete, have been set as shown in Table 5 and Table 6, where the concrete cylinder strength has been estimated as being equal to 85% of the cubic resistances listed in Table 6.

After the PsD testing of the bridges, the tall and the medium piers were tested cyclically until failure (Pinto *et al.*, 1996, Guedes, 1997), respectively up to *230* and *150 mm* of top displacement, under the imposed displacement history shown in Figure 6. Additional cyclic tests (up to *72 mm* at the top of the pier) were carried out on a short pier similar to the one tested in pseudo-dynamic fashion. These cyclic tests on the piers are numerically reproduced herein through a static time-history analysis, so as to enable a first check on the accuracy of the model being assembled. The numerical reproduction of the cyclic test has been performed imposing on the piers the displacement history resulting from the PsD test (Figure 6). In addition, the steel young modulus of the medium-height pier was halved, as suggested by Pinto *et al.* (1996), in order to reproduce the reduction in the stiffness due to the shear damage that this pier suffered prior to this cyclic test (it is recalled that these cyclic tests were carried out after the PsD testing); no reduction in the steel properties was applied to the tall pier, as it was not damaged during the PsD test.

Figure 7 to Figure 9 show a very good match between experimental and numerical results for all the piers; only the reduction in member strength at the very last cycle, when failure occurs, is not perfectly captured.

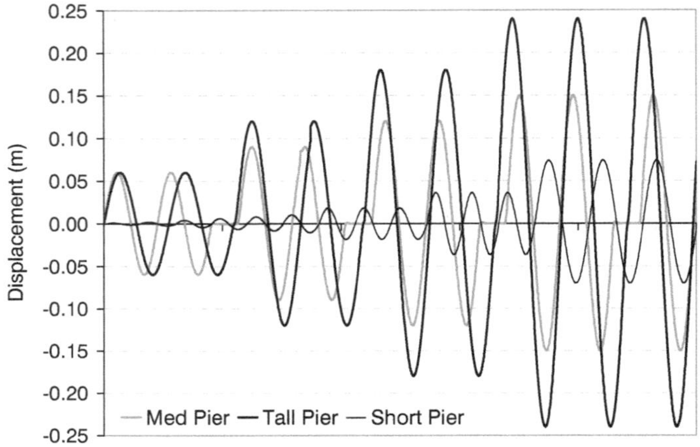

Figure 6. Cyclic test displacement histories

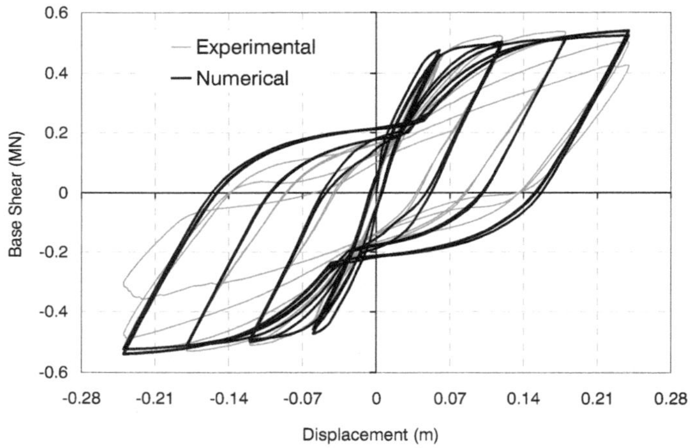

Figure 7. Cyclic Test Results for the Tall Pier

Modelling the bridge deck. Normally, the deck can be modelled as linear elastic, since this is typically the behaviour of real bridges under seismic actions: the deck in fact is generally pre- and/or post-stressed, which means that no damage nor plastic deformations are allowed to occur. Moreover, in the case of isolated bridges, the deck is protected by the isolating system, and it is hardly damaged.

The deck has been modelled with a 3D elastic nonlinear beam-column element, still capable of modelling local geometric nonlinearities. This type of element is fully described by the sectional properties values, based on geometric and mechanical characteristics. In the current

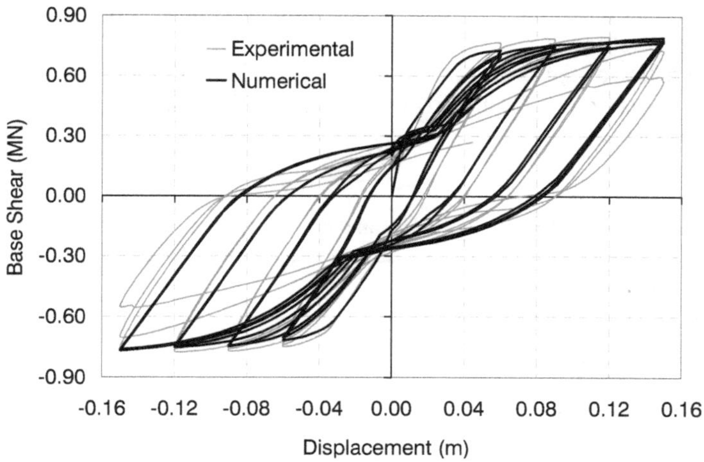

Figure 8. Cyclic test results for the medium pier

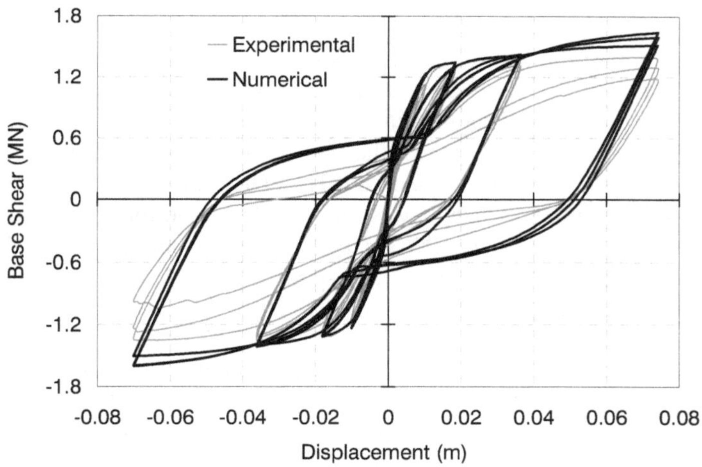

Figure 9. Cyclic test results for the short pier

application, the Young and shear modules have been taken respectively as *25* and *10 GPa*, and the element parameters have been set as listed in Table 3.

The deck has been located at the height of its centre of gravity, *0.602 m* above the pier top, and connected to it by means of a rigid element. The deck can be either modelled as described, or located right at the top of the pier, provided that the moment of inertia with respect to the horizontal axis is translated to that location. In the analysed case this brings to similar results, but the authors opinion is that the first choice is preferable, in order to model more accurately the deck displacement in case of non-rigid connections to the pier, e.g. when a relative rotation of the connection contributes to the drift proportionally to its vertical location.

Before carrying out the nonlinear dynamic analyses, eigenvalue analysis has been run, in order to compare the first transversal modes of the numerical structure (Table 7) with the initial dynamic characteristics of the bridge specimen, computed analytically: the match between test and numerical results is precise for the first mode ($T = 0.183\ s$), and fairly good also for the other two transversal modes.

Table 7. First three modal shapes in the transversal direction of the deck

Mode	Period (s)		Modal shapes
	Test	Analysis	
1st	0.183	0.183	
2nd	0.146	0.148	
3rd	0.085	0.076	

Element Discretization of the FE Model. Each span of the deck has been discretized with four elements, of length equal to *10 %*, *40 %*, *40 %* and *10 %* of the span. The linear elastic behaviour of the element does not strictly call for this fine subdivision, but it has been nonetheless preferred, for sake of accuracy, to refine the mesh near the connections with the piers, where the change of stiffness and properties of the mesh are important. The extremities of the pier constitute the locations for potential plastic hinges, which may be assumed to extend for one twentieth to one tenth of the member length (Priestley *et al.*, 1996), depending on the boundary conditions: as the discretization of the pier pursues the most accurate caption of this phenomenon, each pier has been subdivided into six elements, of length equal to *5 %*, *10 %*, *30 %*, *40 %*, *10 %* and *5 %* of the structural member.

A *FE* analysis requires a careful meshing of the model. The adequate mesh density is achieved when an increase in the number of elements has a negligible effect on the global results: in order to check the effectiveness of the meshing, a second finer discretization was applied, verifying the matching of results in the two cases. It is noted that localization phenomena (i.e. dependence of obtained results on element size) is of reduced relevance here, since the objective is the modelling of the overall response of the bridge.

Other Modelling Details. Rigid connections have been modelled either through elastic frame or link elements: to represent an "infinitely" stiff connection, avoiding numerical difficulties, the stiffness of those elements is set *100* to *1000* times that of adjacent elements. The rigid arm connecting the top of the pier and the centre of gravity of the deck is constituted by an elastic element, whilst the connection between the base of the rigid arm and the top of the pier is modelled as a hinge. The link element representing the latter connects two initially coincident structural nodes and requires the definition of an independent force-displacement (or moment-rotation) response curve for each of its local six DOF: in order to model the engaging of the sub and superstructure of the pier-deck connection in the transversal direction by mean of shear keys, the link element is set as a spring with infinite stiffness in the vertical and transversal direction, and fully flexible in all the other four DOF. Boundary conditions are defined as restraints in global coordinates: the deck is simply supported at the first end and hinged at the other, whilst piers are fully fixed at the base. In Figure 10, the above described connection details are depicted.

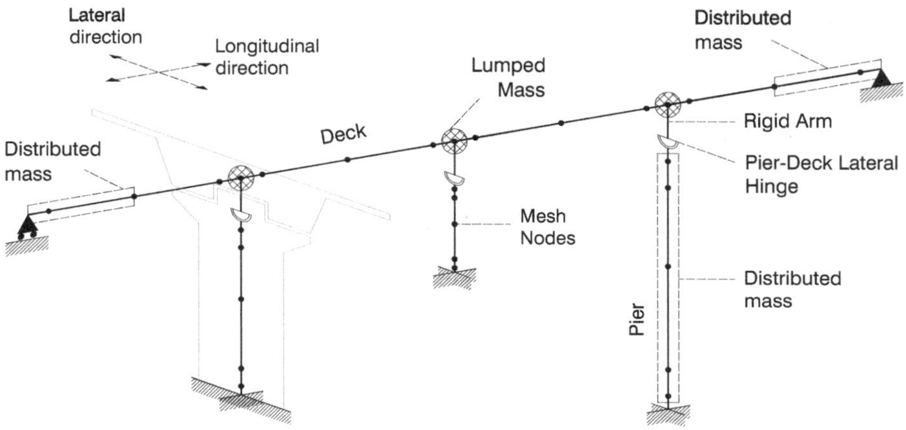

Figure 10. Details of the FE model of the bridge

Modal and dynamic analyses require as a matter of necessity the definition of the masses, which can be either lumped nodal masses or distributed. The piers were characterized with a distributed mass element of *1.664 ton/m*, whilst the deck mass was concentrated at the top of each pier for an amount corresponding to the tributary deck length (*56 tons* per pier), and distributed at the two half span deck extremities with an amount of *2.784 ton/m*.

Equivalent viscous damping, with values typically ranging around 1 ÷ 2%, is customarily introduced in order to somehow represent minor energy dissipation mechanism other from the hysteretic ones (e.g. friction across cracks, radiation through foundations, and so on). This usually involves the employment of Rayleigh damping matrices, whereby damping is defined as proportional to the mass and stiffness of the structural members (see Clough and Penzien, 1994). However, given the uncertainties associated to the quantification of such equivalent viscous damping, and considering also the recent doubts raised with regards to the employment of Rayleigh damping in nonlinear dynamic analyses (Hall, 2005, Priestley and Grant, 2005), this minor source of dissipation has been conservatively neglected in the numerical simulations.

A static load of *1700 kN* has been applied at the top of each pier, representing the deck weight according to the test setup, whilst the time history described in Figure 4, including the *10 s* interval with no acceleration (needed to damp out the structure motion after the first earthquake run), has been imposed at the pier bases and at the abutments. In this manner, the cumulative damage effects caused by the testing of the same structure under to successive earthquake input motions are adequately modelled.

The time step for the dynamic analysis has been selected as *0.004 s*, coincident with the input record sampling time step (hence the input motion is accurately considered), and sufficiently small with respect to the dominant vibration period of the structure (*0.4 secs*), so as to guarantee numerical stability. For what concerns the nonlinear solution algorithm, the Hilber-Hughes-Taylor (1977) integration scheme was employed, associated to a displacement and force based convergence criterion.

5.3 Comparisons Between Numerical and Experimental Results

The current section presents the comparisons between numerical and experimental results, in terms of displacements and forces at the top of the short, medium and tall piers observed when the bridge was subjected to the second and stronger earthquake input motion. Figure 11 to Figure 13 show results of the top displacement (left) and the top shear (right): there is a good agreement, in terms of both the amplitude and the frequency content of the response. Table 8 illustrates the ratios of the maximum absolute response obtained from numerical calculation to that from tests. It is noted that the force response of the squat pier is not reproduced with full accuracy, whereas displacements are instead very well predicted. The numerical overestimation of the action at the top of the short pier can be explained by the fact that the fibre-based element formulation employed did not feature the possibility of modelling shear flexibility (or section torsion/warping): the fact that the stiffness of this pier is not reduced, as it would be due to the shear damage.

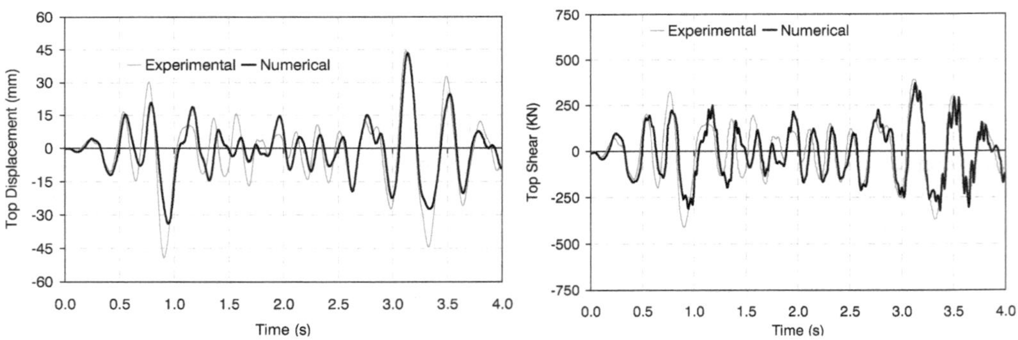

Figure 11. Tall pier top displacements (left) and top shear (right)

Table 8. Ratios of the absolute maximum response obtained from numerical calculation to that from tests

	Tall Pier	Med Pier	Short Pier
Displacement	88%	94%	102%
Top Shear	90%	95%	188%

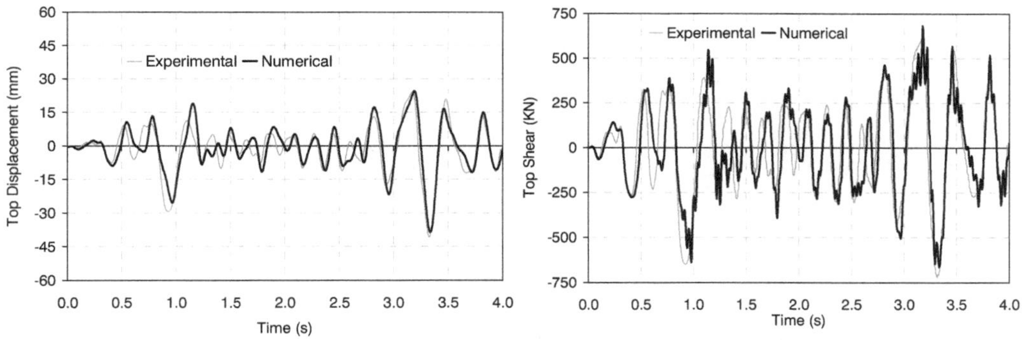

Figure 12. Medium pier top displacements (left) and top shear (right)

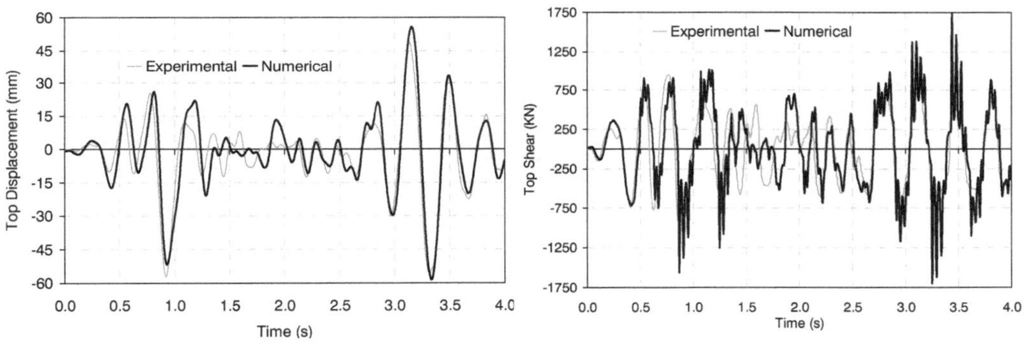

Figure 13. Short pier top displacements (left) and top shear (right)

6 Case Study 2: Multi-storey RC Building

6.1 Description of the model

The second case study concerns the dynamic analysis of a multi-storey RC building. A four-storey, three-bay reinforced concrete bare frame, Figure 14, was designed (Carvalho *et al.*, 1999) and built at the European Laboratory for Structural Assessment (ELSA) of the Joint Research Centre (JRC) at Ispra, Italy (Pinto *et al.*, 2002). The full-scale model was constructed for pseudo-dynamic testing, under the auspices of the EU-funded ECOEST/ICONS programme. The frame was designed essentially for gravity loads and a nominal lateral load of *8 %* of its weight. The reinforcement details attempted to mirror the construction practice of southern European countries in the 1950's and 1960's. A detailed description of the analytical tool, the modelling approach as well as the nonlinear dynamic analyses used is presented herein.

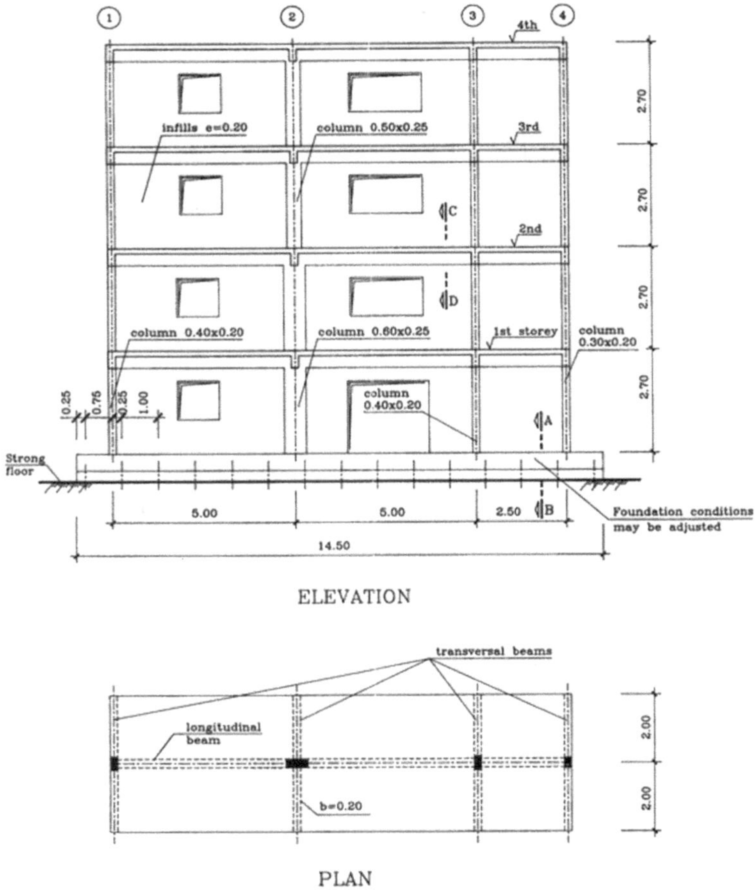

Figure 14. Elevation and plan views of the frame, after Carvalho *et al.* (1999)

The four-storey bare frame examined in this section is known to have a soft storey at the third floor. This is attributed to a drastic change in strength and stiffness at this level through a reduction in both the reinforcement content and the section dimensions in the columns between the second and third storeys, coinciding also with the location of lap-splicing. Such characteristics are common in buildings designed predominantly for gravity-loading and the failure of a storey midway up a building has been observed in past devastating earthquakes such as the Kobe earthquake of 1995 (e.g. EERI, 1997).

The columns were non-ductile, smooth reinforcing bars were used, capacity design principles were ignored and lap splicing occurred in critical regions (Figure 15). Detailed information on the frame and the set up of the experimental test can be found in Pinto *et al.* (2002) and Pinho and Elnashai (2000).

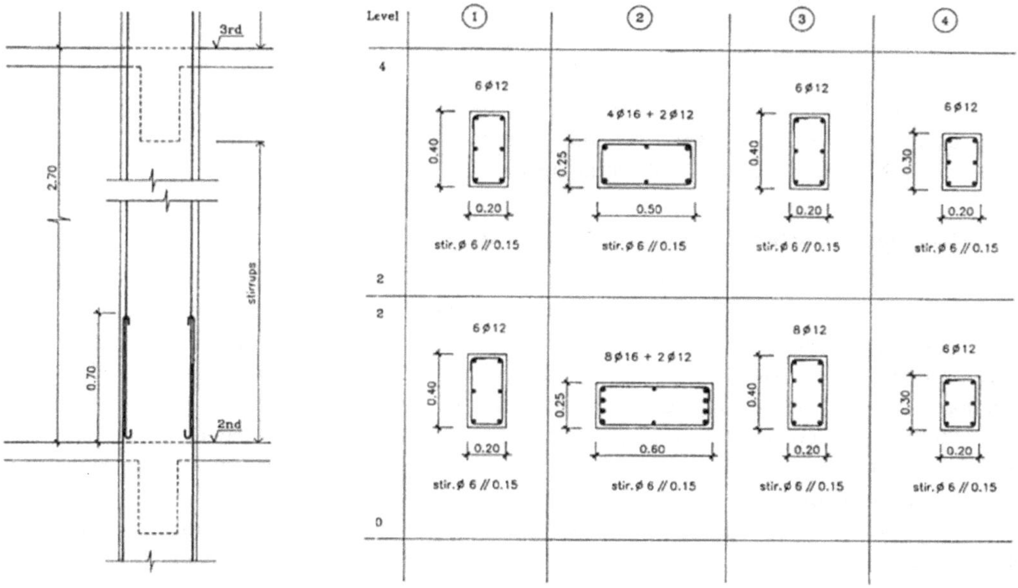

Figure 15. Reinforcement detail of the columns, after Carvalho *et al.* (1999)

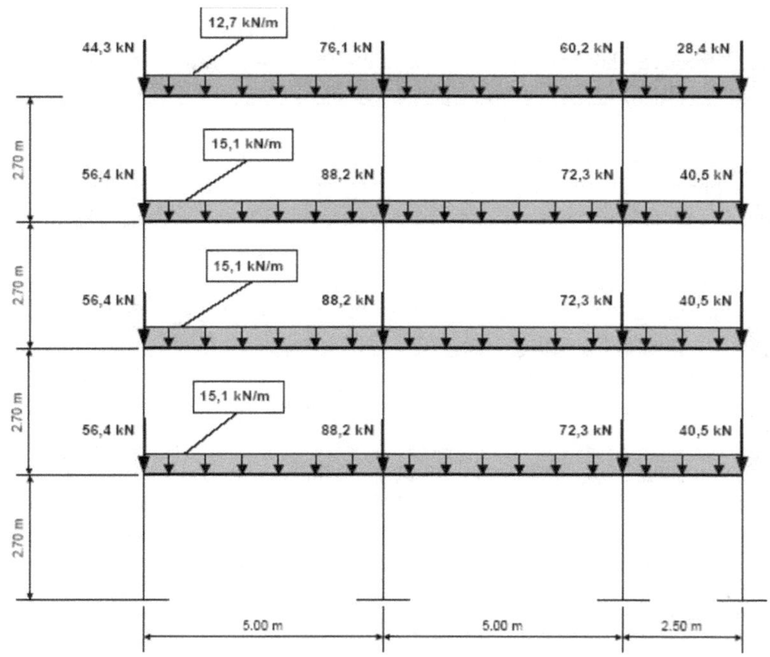

Figure 16. Scheme of vertical loads for nonlinear analysis, after Carvalho *et al.* (1999).

The materials considered at the design phase were a low strength concrete class *C16/20* (CEN, 1991) and smooth reinforcement steel class *Fe B22k* (Italian standards). The latter refers to smooth bars with a yield stress of *235 MPa* and ultimate strength of *365 MPa*.

The vertical loads considered in the design consisted of the self-weight of the slab and transverse beams, finishes, infill walls and the quasi-permanent static load. Figure 16 shows the scheme of vertical loads applied to the structure.

6.2 Modelling of the Structure in the FE Program

As in the previous case-study, the finite element analysis program SeismoStruct (SeismoSoft, 2005) is employed to run all analyses. Structural members have been discretised through the use of beam-column elements, which are recalled to take account of geometrical nonlinearity and material inelasticity. Also as discussed before, since a constant generalized axial strain shape function is assumed in the adopted cubic formulation of the element, it results that its application is only fully valid to model the nonlinear response of relatively short members and hence a number of elements (usually three to four per structural member) is required to accurate model the structural frame members.

In total, *112* inelastic frame elements, capable of representing progressive cracking and spread of inelasticity, are used to model the RC frame. All member sections are represented explicitly and, for the purpose of strain/stress evaluation, are subdivided into a number of fibres ($200 \div 300$) varying according to the section size. The length of the elements varies according to their location, with smaller elements being used in the vicinity of beam-column connections where large levels of inelastic deformation are expected.

All connections are assumed rigid and fully-fixed boundary conditions are adopted at the base of the building. Effective slab widths of *1.0 m* and *0.65 m* were adopted for the long and short spans, respectively, following the formulae presented in Eurocode 2 [CEN, 1991].

Vertical loads and masses are applied at each beam node and at the beam-column joints mirroring the load and mass distribution presented previously in Figure 16.

Concrete is represented by a uniaxial constant confinement model (Mander *et al.*, 1988) and is calibrated using the concrete characteristics values obtained during testing. As for the reinforcement bars, the Menegotto-Pinto (1973) steel model, with an isotropic hardening constitutive relationship (Filippou *et al.*, 1983), was adopted.

It is worth mentioning that shear strains across the element cross section are not modelled; in addition, warping strains and warping effects are not considered in the current formulation, either. Additionally, the elastic torsional rigidity is used in the formulation of the nonlinear frame elements; this clearly involves some degree of approximation for the case of reinforced concrete sections.

No viscous damping was considered in any dynamic analysis, since energy dissipation through hysteresis is already implicitly included within the nonlinear fibre model formulation of the inelastic frame elements, and non-hysteretic type damping was assumed to be negligible within the scope of the present endeavour (see also previous discussion on this issue).

6.3 Nonlinear Dynamic Analysis

Earthquake input. A collection of artificial records was available for use in the pseudo-dynamic experimental programme. The records were derived following a probabilistic seismic hazard analysis carried out by Campos-Costa and Pinto (1999). For the purpose of the experimental programme, the severity class "Moderate-High", typical of Southern European countries, was chosen. A set of hazard-consistent time histories was artificially generated to fit the uniform risk spectra (URS) for return periods of *100*, *475*, *975* and *2000* years.

In Figure 17, the acceleration and displacement elastic response spectra for all accelerograms, computed for an equivalent viscous damping of *5 %*, are shown. These indicate peak acceleration response for periods of vibration of up to *0.5* seconds, with values ranging from *0.3 g* to *1.1g*. Figure 18 shows the artificial Acc-475 (*475* years return period) and Acc-975 (*975* years return period) accelerograms; only these two records will be considered in the nonlinear dynamic and pushover analyses carried out herein. As is common with artificial records, a wide range of frequencies is present in the accelerograms.

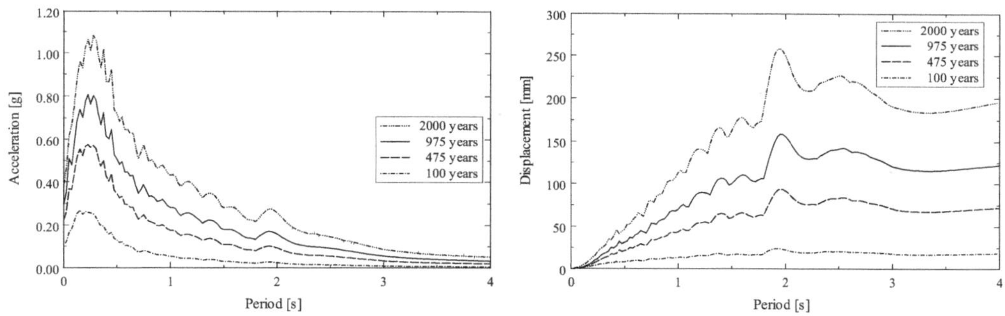

(a) (b)

Figure 17. Response spectra of input motion: (a) acceleration; (b) displacement.

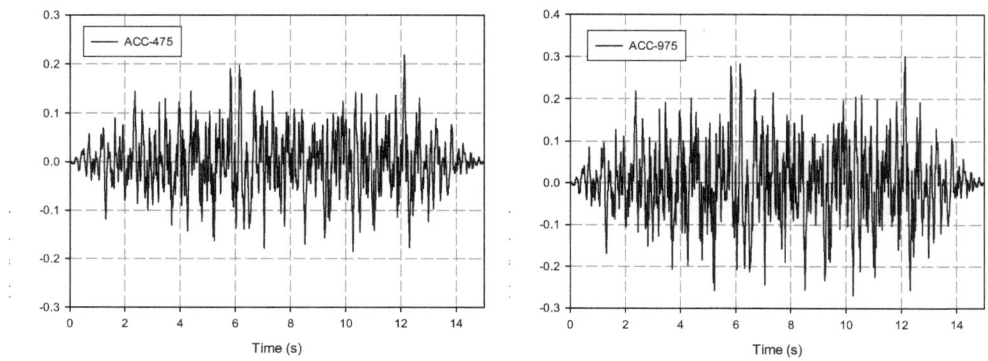

(a) (b)

Figure 18. Artificial acceleration time histories for
(a) 475 year (Acc-475) and (b) 975 year (Acc-975) return period.

Comparison of experimental and analytical results. The bare frame specimen of Figure 14 was subjected to a pseudo-dynamic test using the Acc-475 input motion and subsequently to a second test carried out with the Acc-975 input motion. The results of the tests showed that the deformation demand concentrated in the 3rd storey for the Acc-475 test and that collapse of the 3rd storey was almost reached for the Acc-975 test (Pinho and Elnashai, 2000).

Nonlinear dynamic analyses of the model have been carried out for the two records by placing them consecutively in order to reproduce the testing sequence and better predict the behaviour of the frame. The top frame displacements, obtained for the experimental test, for the Acc-475 record and the Acc-975 record are shown in Figure 19a and Figure 19b, respectively. In addition, the top displacement response obtained by numerical simulation of each test has been superimposed. It may be concluded that the analytical prediction agrees with the results obtained through testing.

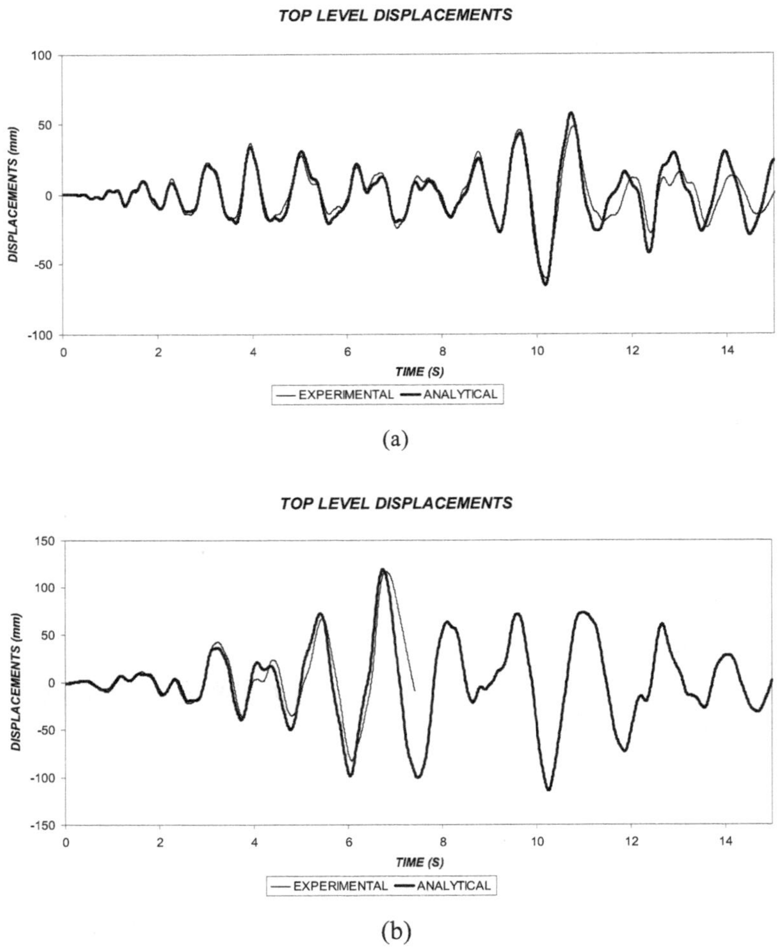

Figure 19. Analytical and experimental top frame displacement: (a) Acc-475; (b) Acc-975

Inter-storey drift is a crucial parameter in terms of structural response since it is closely related to the damage sustained by buildings during seismic action. The soft-storey in the test case frame can only be identified through observation of the inter-storey drift profile. Figure 20 shows the drift profiles at the peak displacement from both the analytical and experimental time-histories; it is clear that the analytical model is able to predict the soft-storey at the third floor. Further refinement of the analytical model could perhaps produce a closer match between the analytical and experimental drift profiles but this is not within the scope of the present exercise. It suffices that the nonlinear fibre analysis can predict the soft-storey at the third floor and thus this will be the reference to which all other analytical analyses will be compared. Note that no post-test calibration has been carried out. The structure was modelled as it is. The differences between analysis versus experiment are likely to be due to the fact the third storey developed a shear failure mechanism, not yet incorporated in the program used.

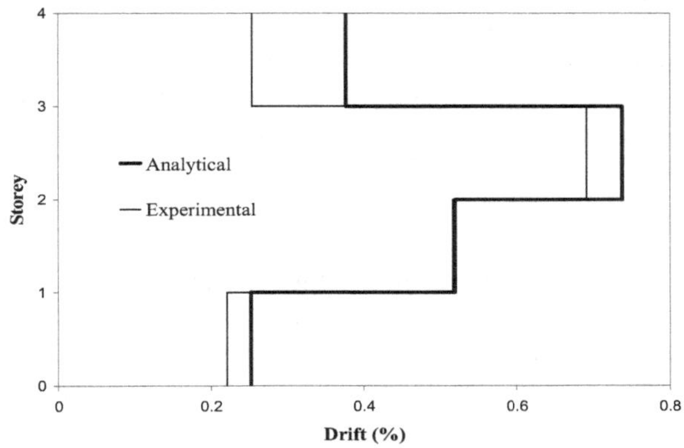

(a)

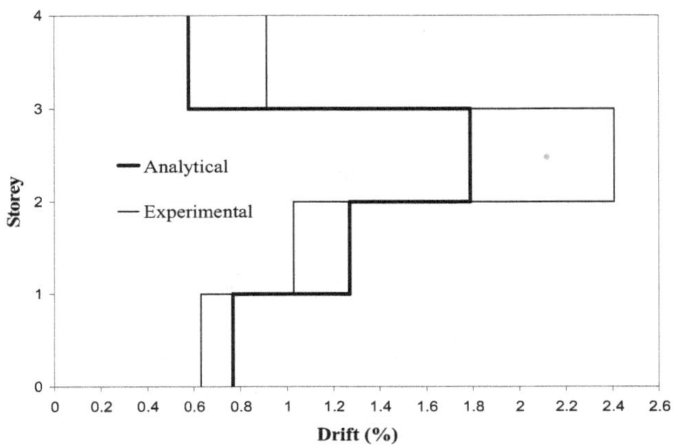

(b)

Figure 20. Analytical and experimental drift profiles: (a) Acc-475; (b) Acc-975

7 Conclusions

Structural behaviour is inherently nonlinear, particularly in the presence of large displacements or material nonlinearities, the structural response can be accurately caught only by means of nonlinear dynamic analyses. The fibre modelling approach employed in the current work is shown to be capable of associating simplicity of use, even for not highly experienced users. Moreover, its ability to simulate the nonlinear dynamic response of reinforced concrete bridges or buildings to seismic loads has been proven by simulating large-scale experimental pseudo-dynamic tests. Results of the dynamic and modal analyses performed reveal a good agreement with the pseudo-dynamic tests, both in terms of displacements and forces at the top of the tall and medium-height piers. At present, shear strains across the element cross-section are not included in the fibre-element formulation adopted, i.e. the strain state of a section is fully represented by the curvature at centroidal axial strains alone: this approach is not accurate enough for representing the squat pier deformation state, where shear deformations are of relevance.

In closing, simple-to-calibrate fibre structural models can be employed to reproduce with good level of accuracy the nonlinear structural response of continuous span bridge structures and multi-storey RC buildings. In other words, it is believed that such an advanced analytical tool can be readily handled within a common engineering practise framework, provided a basic level of awareness on the decisions that the designer has to face, discussed in this paper, is available.

Acknowledgements

The author would like to acknowledge the important contribution of Dr C. Casarotti and Dr M. López-Menjivar for providing much of the material included in this manuscript. The assistance of Dr A. Pinto, Dr J. Guedes and Dr P. Delgado in kindly providing the experimental data and the information required to run the analyses described in this work is also gratefully acknowledged. The author would also like to thank Dr S. Antoniou for the valuable assistance in the development and use of the *FE* code employed in the current endeavour. Finally, the assistance of Mr D. Bolognini in the writing-up of this document is gratefully acknowledged.

References

Banon, H., Biggs, J., and Irvine, M., (1981). Seismic Damage in Reinforced Concrete Frames, *J. Struct. Engrg.*, ASCE, 107 (ST9), 1713-1729.

Bazant, S., and Bhat, P., (1977). Prediction of Hysteresis in Reinforced Concrete Members, *J. of Struc. Engrg.*, ASCE, 103 (ST1), 151-167.

Bertero, V., Aktan, A., Charney, F., and Sause, R., (1984). Earthquake Simulator Tests and Associated Experimental, Analytical and Correlation Studies of One-Fifth Scale Model, *in Earthquake Effects on Reinforced Concrete Structures*, American Concrete Institute, SP-84-13, Detroit, 375-424.

Brancaleoni, F., Ciampi, V. , and Di Antonio, R., (1983). Rate-Type Models for Non Linear Hysteretic Structural Behavior, *EUROMECH Colloquium*, Palermo, Italy.

Carvalho, E. C., Coelho, E., and Campos-Costa, A., (1999). Preparation of the full-scale tests on reinforced concrete frames. Characteristics of the test specimens, materials and testing conditions, *ICONS Report*, Innovative Seismic Design Concepts for New and Existing Structures, European TMR Network, LNEC.

Campos-Costa, A., and Pinto, A.V., (1999). *European Seismic Hazard Scenarios – An Approach to the Definition of Input Motions for Testing and Reliability Assessment of Civil Engineering Structures*, JRC Special Pubblication, no X.99.XX, ELSA, JRC – Ispra, Italy.

Comité Europeen de Normalization, (2002). Eurocode 8: Design of Structures for Earthquake Resistance - Part 2: Bridges, PrEN 1998-2: 2003, 2 April 2002, CEN, Brussels, Belgium

Charney, F., and Bertero, VV., (1982). An Evaluation of the Design and Analytical Seismic Response of a Seven Story Reinforced Concrete Frame-Wall Structure. *EERC Report 82/08*, Earthquake Engrg. Research Center, Univ. of California, Berkeley.

Ciampi, V., and Carlesimo, L., (1986). A Nonlinear Beam element for Seismic Analysis of Structures, *8th European Conf. Earthquake Engrg.*, Lisbon.

Clough, R., and Benuska, L., (1967). Nonlinear Earthquake Behavior of Tall Buildings, *J. Engrg. Mech.*, ASCE, 93(EM 3), 129-146.

Clough, R. W., and Penzien , J., (1994). *Dynamics of Structures*, 2nd Edition, McGraw Hill.

Clough, R., and Johnston, S., (1966). Effect of Stiffness Degradation on Earthquake Ductility Requirements, *Transactions of Japan Earthquake Engineering Symposium*, Tokyo, 195-198.

Cook, D., Malkus, D. S., and Plesha, M. E., (1989). *Concepts and applications of Finite Element Analysis*, 3rd Edition, Wiley & Sons, Inc.

Crisfield, M. A., (1997). *Non-linear Finite Element Analysis of Solids and Structures, Volume 2 - Advanced Topics*, John Wiley and Sons Ltd.

Earthquake Engineering Research Institute (EERI), (1997). Recovery and Reconstruction from Recent Earthquakes, *Proc. 5th United States / Japan Workshop on Urban Earthquake Hazard Reduction*, Pasadena, CA, United States, pp. 455

Filippou, F. C., Popov, E. P., and Bertero, V. V., (1983). Modelling of R/C joints under cyclic excitations, *J. Struct. Engrg.*, 109(11), 2666-2684.

Filippou, F. C., and Issa, A., (1988). Nonlinear Analysis of Reinforced Concrete Frames Under Cyclic Load Reversals, *EERC Report 88/12*, Earthquake Engrg. Research Center, Univ. of California, Berkeley.

Fragiadakis, M., Pinho, R., and Antoniou, S., (2006). Modelling Inelastic Buckling of Reinforcing Bars under Earthquake Loading, *J. Struct. Engrg.*, ASCE. (submitted for publication)

Giberson, M., (1967). The Response of Nonlinear Multi-Story Structures Subjected to Earthquake Excitations, *Earthquake Engineering Research Laboratory*, Pasadena.

Guedes, J., (1997). Seismic behaviour of reinforced concrete bridges. Modelling, numerical analysis and experimental assessment, *PhD Thesis*, Univ. of Porto, Porto, Portugal.

Hall, J. F., (2005). Problems encountered from the use (or misuse) of Rayleigh damping, *Earthquake Engrg and Struct. Dyn.*, 35(5), 525-545.

Hellesland, J. and, Scordelis, A., (1981). Analysis of RC Bridge Columns Under Imposed Deformations, *IABSE Colloquium*, Delft, 545-559.

Hilber H. M., Hughes T. J. R., and Taylor R. L., (1977). Improved numerical dissipation for time integration algorithms in structural dynamics, *Earthquake Engrg. and Struct. Dynamics*, 5(3), 283-292.

Izzuddin, B. A., (2001). Conceptual issues in geometrically nonlinear analysis of 3D framed structures, *Comp. Meth. Applied Mech. and Engrg.*, 191, 1029-1053.

Kaba, S., and Mahin, S. A., (1984). Refined Modeling of Reinforced Concrete Columns for Seismic Analysis, *EERC Report 84/03*, Earthquake Engrg. Research Center, University of California, Berkeley.

Kachlakev, D., Miller, T., Yim, S., Chansawat, K., and Potisuk, T., (2001). Finite element modeling of reinforced concrete structures strengthened with frp laminates, *Final Report - SPR 316*, Oregon Department of Transportation Research Group.

Mahasuverachai, M., and Powell, G. H., (1982). Inelastic Analysis of Piping and Tubular Structures, *EERC Report 82/27*, Earthquake Engrg. Research Center, University of California, Berkeley.

Mander, J. B., Priestley, M. J. N., and Park, R., (1988). Theoretical Stress-Strain Model for Confined Concrete, *J. of Structural Engrg.*, ASCE, 114 (ST8), 1804-1826.

Mari, A., and Scordelis, A., (1984). Nonlinear Geometric Material and Time Dependent Analysis of Three Dimensional Reinforced and Prestressed Concrete Frames, *SESM Report 82-12*, Dept. Civil Engrg, University of California, Berkeley.

Martinez-Rueda, J. E., and Elnashai, A. S., (1997). Confined concrete model under cyclic load, *Materials and Structures*, 30(197), 139-147.

Menegotto, M., and Pinto P. E., (1973). Method of analysis for cyclically loaded R.C. plane frames including changes in geometry and non-elastic behaviour of elements under combined normal force and bending, *Symposium on the Resistance and Ultimate Deformability of Structures Acted on by Well Defined Repeated Loads*, International Association for Bridge and Structural Engineering, Zurich, Switzerland, 15-22.

Monti, G., Nuti, C., (1992). Nonlinear cyclic behaviour of reinforcing bars including buckling, *J. Struct. Engrg*, 118(12), 3268-3284.

Newmark, N. M., (1959). A method of computation for structural dynamics, *J. of the Engrg. Mechanics Division*, ASCE, Vol. 85, No. EM3, 67-94.

Otani, S., (1974). Inelastic Analysis of R/C Frame Structures, *J. of Struct. Engrg.*, ASCE, 100 (ST7), 1433-1449.

Papaioannou, I., Fragiadakis, M., and Papadrakakis, M., (2005). Inelastic analysis of framed structures using the fiber approach, *Proc. 5th GRACM International Congress on Computational Mechanics*, Limassol, Cyprus.

Pinho, R., (2000). Shaking Table Testing Of RC Walls, *ISET J. of Earthquake Technology*, 37(4), 119-142.

Pinho, R., and Elnashai, A. S., (2000). Dynamic collapse testing of a full-scale four storey RC frame, *ISET Journal of Earthquake Technology*, Vol. 37, no. 4, 143-163.

Pinto, A. V., Verzeletti, G., Pegon, P., Magonette, G., Negro, P., and Guedes, J., (1996). *Pseudo-Dynamic Testing of Large-Scale R/C Bridges*, Research Report EUR 16378 EN, Joint Research Centre, Ispra, Italy.

Pinto, A. V., Verzeletti, G., Molina, J., Varum, H., Pinho, R., and Coelho, E., (2002). *Pseudo-dynamic tests on non-seismic resisting RC frames (bare and selective retrofit frames)*, Research Report EUR 20244 EN, Joint Research Centre, Ispra, Italy.

Priestley, M. J. N., Seible, F., and Calvi, G. M., (1996). *Seismic Design and Retrofit of Bridges*, John Wiley & Sons Inc., New York

Priestley, M. J . N., and Grant, D. N., (2005). Viscous damping in seismic design and analysis, *J. of Earthquake Engrg.*, 9 (Special Issue 2), 229-255.

Scott, B. D., Park, R., and Priestley, M. J. N., (1982). Stress-Strain Behavior of Concrete Confined by Overlapping Hoops at Low and High Strain Rates, *ACI Journal*, 79(1), 13-27.

SeismoSoft, (2005). SeismoStruct - A Computer Program for Static and Dynamic nonlinear analysis of framed structures [on line], available from URL: http//www.seismosoft.com (March 30, 2005).

Spacone, E., (1994). Flexibility-Based Finite Element Models for the Nonlinear Static and Dynamic Analysis of Concrete Frame Structures, *Ph.D. Dissertation.*, Department of Civil Engrg., Univ. of California, Berkeley.

Spacone, E., (2001). A module for Analysis and Design of Segmental Prestressed Concrete Bridges (CASI-TR-01-04), Final report of a CASI FY00 Technology Transfer Grant. Fort Collins, CO: Colorado Advanced Software Institute.

Sullivan, T., Pinho, R., and Pavese, A., (2004). *An Introduction to Structural Testing Techniques in Earthquake Engineering*, Educational Report ROSE 2004/01, IUSS Press, Pavia, Italy.

Takeda, T., Sozen, M. A., and Nielsen, N., (1970). Reinforced Concrete Response to Simulated Earthquakes, *J. of Struct. Engrg.*, ASCE, 96 (ST12), 2557-2573.

Takizawa, H., (1976). Notes on Some Basic Problems in Inelastic Analysis of Planar RC Structures, *Transaction of Architecture Institute of Japan*, 240, Part I in February, 51-62, Part II in March, 65-77.

Zeris, C. A., and Mahin, S. A., (1988). Analysis of Reinforced Concrete Beam-Columns Under Uniaxial Excitation, *J. of Structural Engrg.*, ASCE, 114 (ST4), 804-820.

Zeris, C. A., and Mahin, S. A., (1991). Behavior of Reinforced Concrete Structures Subjected to Biaxial Excitation, *J. of Structural Engrg.*, ASCE, 117 (ST9), 2657-2673.

Using Pushover Analysis for Assessment of Buildings and Bridges

Rui Pinho

Department of Structural Mechanics, University of Pavia, Pavia, Italy

Abstract. Estimating seismic demands on structures requires explicit consideration of the structural inelastic behaviour: to this end, the use of nonlinear static procedures, or pushover analyses, is inevitably going to be favoured over more complex nonlinear time-history analysis methods. Currently employed pushover methods are performed subjecting the structure to monotonically increasing lateral forces with invariant distribution until a target displacement is reached, basing both the force distribution and target displacement on the assumptions that the response is controlled by the fundamental mode, unchanged after the structure yields. However, these invariant force distributions cannot account for the contributions of higher modes to response, nor for the redistribution of inertia forces because of structural yielding and the associated changes in the vibration properties: in order to overcome drawbacks arising from conventional methods; an innovative displacement-based adaptive pushover technique for estimation of the seismic capacity of RC structures is illustrated. Analytical parametric studies show that, with respect to conventional pushover methods, the novel approach can lead to the attainment of significantly improved predictions, which match very closely results from dynamic nonlinear analysis.

1 Introduction

It is unquestionable that nonlinear dynamic analysis is the most accurate method for assessing the response of structures subjected to earthquake action. Indeed, any type of static analysis will always be inherently flawed, given the conspicuous absence of time-dependent effects. However, as noted by Goel and Chopra (2005a), amongst others, such type of analysis is not without its difficulties or drawbacks, particularly for what concerns application within a design office environment.

Firstly, in order to employ dynamic analysis for seismic design/assessment of structures, an ensemble of site-specific ground motions compatible with the seismic hazard spectrum for the site must be simulated. As described by Bommer and Acevedo (2004), amongst others, this is, however, a far from simple task, since seismic design codes feature insufficient or inadequate guidance on procedures to either (i) generate artificial spectrum-compatible records, (ii) produce synthetic accelerograms from seismological models or (iii) select appropriate suites of real acceleration time-series, eventually modified to better fit a given code response spectrum. It is believed that until better guidance on record selection/generation will be made available to earthquake engineer designers, this first step will remain as a very difficult-to-overcome hurdle to the use of dynamic time-history analysis in design office applications.

Secondly, notwithstanding the significant increase in computing power witnessed in recent years, nonlinear time-history analysis remains computationally demanding, especially when fibre-based (distributed inelasticity) structural analysis programs, which are simpler to calibrate than their plastic-hinge (concentrated plasticity) counterparts, are employed to model the seismic response of large multi-storey irregular buildings, requiring 3D models with thousands of elements. This problem becomes even the more significant if one considers that the analyses will need to be repeated a significant amount of times, not only because design codes or guidance documents request for a relatively large number of earthquake records to be employed in order to warrant minimum probabilistic validity of the results, but also, and perhaps mainly, because the process of analysing any given structure is invariably an iterative one, given that modelling errors are commonly encountered as the design/assessment process evolves.

Thirdly, even in those situations where the expertise and resources for running time-history analyses are available, it is often the case that preliminary simpler analysis (i.e. modal and static analyses) are run to enable a first check of the model; errors in the definition/assemblage of a finite elements model are difficult to detect from dynamic analysis results, whilst they tend to be relatively evident from the output of eigenvalue or pushover runs. As an example, inspection of the first modes of vibration of a given building model may be used to check if member stiffness has been correctly allocated or if the mass has been appropriately distributed, whilst examination of a force-displacement monotonic capacity curve may serve to quickly assess if member strength and ductility has been properly assigned. Static analyses, even if representing simplified methods, also provide important structural response information such as (i) identification of critical regions, where large inelastic deformations may occur, (ii) individuation of strength irregularities in plan and elevation that might cause important changes in the inelastic dynamic response characteristics, (iii) evaluation of the force demand in potentially brittle elements, and (iv) prediction of the sequence of yielding and/or failure of structural members. In addition, the explicit insight that pushover-derived base shear vs. top displacement capacity curves provide into the stiffness, strength and ductility of a given structure, constitutes the type of qualitative data that is always most informative and useful within a design application, even when time-history analysis is then employed for the definitive verifications.

The above constitute, in the opinion of the author, strong reasons for nonlinear static analysis methods to continue to be developed and improved, so that these tools can become even more reliable and useful when employed either as a replacement to time-history analysis in the seismic design/assessment of relatively simple non-critical structures, or as a complement to dynamic analysis of more complex/critical facilities.

2 Definition and Scope

The term pushover analysis describes a modern variation of the classical collapse analysis method, as fittingly described by Kunnath (2004). It refers to an analysis procedure whereby an incremental-iterative solution of the static equilibrium equations is carried out to obtain the response of a structure subjected to monotonically increasing lateral load pattern. The structural resistance is evaluated and the stiffness matrix is updated at each increment of the forcing function, up to convergence. The solution proceeds until (i) a predefined performance limit state is reached, (ii) structural collapse is incipient or (iii) the program fails to converge. In this manner, each point in the resulting displacement vs. base shear capacity curve represents an effective and equilibrated stress-state of the structure, i.e. a state of deformation that bears a direct correspondence to the applied external force vector.

Within the framework of earthquake engineering, pushover analysis is employed with the objective of deriving, with relative ease, an envelope of the response parameters that would otherwise be obtained through a much more complex and time-consuming Incremental Dynamic Analysis (IDA) procedure, as can be construed by Figure 1. IDA is a parametric analysis method by which a structure is subjected to a series of nonlinear time-history analyses of increasing intensity (Vamvatsikos and Cornell, 2002), with the objective of attaining an accurate indication of the "true" dynamic response of a structure subjected to earthquake action.

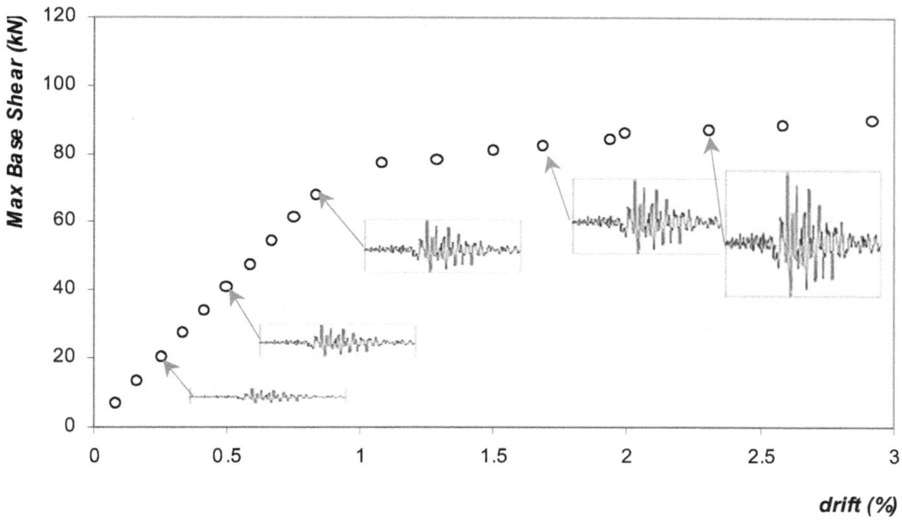

Figure 1. Maximum base-shear and top displacement values obtained with incremental dynamic analysis.

Recent years have also witnessed the development and introduction of an alternative type of nonlinear static analysis, which involves running multiple pushover analyses separately, each of which corresponding to a given modal distribution, and then estimating the structural response by combining the action effects derived from each of the modal responses (i.e. each displacement-force pair derived from such procedures does not actually correspond to an equilibrated structural stress state). Paret *et al.* (1996) first suggested the Multi-Modal Pushover procedure, which was then refined by Moghadam and Tso (2002). Chopra and Goel (2002), on the other hand, have developed and proposed a Modal Pushover Analysis (MPA) technique, which Hernández-Montes *et al.* (2004) have then adapted into an Energy-based Pushover formulation. A further refinement of such multiple-pushover procedures, with the aim to account for the alteration of local resistance and modal characteristics of the structure induced by the accumulation of damage, consists in the employment of adaptive updating of the loading pattern (Gupta and Kunnath, 2000; Aydinoglu, 2003), effectively meaning that the methods may now be categorised as piecewise linear response spectrum analysis. As highlighted by their respective authors, the main advantage of this category of static analysis procedures is that they may be applied using standard readily-available commercial software packages. The associated drawback, however, is that the methods are inevitably more complex than running a single pushover analysis, as noted by Maison (2005). Furthermore, some of the proposed "multiple-run" procedures, either they have an adaptive or non-adaptive nature, lead to difficulties when applied within capacity-spectrum type of procedure due to the need to handle capacity curves associated with higher mode force patterns that display a reversal of the roof displacement as inelasticity develops in the structure (Hernàndez-Montes *et al.* 2004; ATC 2005; Goel and Chopra 2005b). For all of the above, these multiple-pushover based approaches do not constitute the scope of the current presentation, where focus is placed instead on "single-run" pushover analysis procedures.

In tandem with the present drive for performance-based seismic engineering, there is also currently a thrust for the development and code implementation of displacement or, more generally, deformation-based design and assessment methods. Therefore, it would seem that applying displacement loading, rather than force actions, in pushover procedures would be an appropriate option for nonlinear static analysis of structures subjected to earthquake action. However, due to the unvarying nature of the applied displacement loading vector, conventional (non-adaptive) displacement-based pushover analysis can conceal important structural characteristics, such as strength irregularities and soft storeys, should the displacement pattern adopted at the start of the analysis not correspond to the structure's post-yield failure mechanism. Consequently, when only non-adaptive static nonlinear analysis tools are available, as has been the case throughout the past, force-based pushover does constitute a preferable choice over its displacement-based counterpart.

On the other hand, however, if one is able to apply displacements, rather than forces, in an adaptive fashion, that is, with the possibility of updating the displacement loading pattern according to the structural properties of the model at each step of the analysis, then a conceptually appealing deformation-based nonlinear static analysis tool is obtained.

3 Recent Developments in Single-Run Pushover Analysis

According to recently introduced code provisions, such as FEMA-356 (BSSC, 2000) and Eurocode 8 (CEN, 2002), pushover analysis should consist of subjecting the structure to an increasing vector of horizontal forces with invariant pattern. Both the force distribution and target displacement are based on the assumptions that the response is controlled by the fundamental mode and the mode shape remains unchanged until collapse occurs. Two lateral load patterns, namely the first mode proportional and the uniform, are recommended to approximately bound the likely distribution of the inertia forces in the elastic and inelastic range, respectively.

However, a number of recent studies, summarised in the FEMA-440 report (ATC, 2005), raise doubts on the effectiveness of these conventional force-based pushover methods in estimating the seismic demand throughout the full deformation range: (i) inaccurate prediction of deformations when higher modes are important and/or the structure is highly pushed into its nonlinear post-yield range, (ii) inaccurate prediction of local damage concentrations, responsible for changing the modal response, (iii) inability of reproducing peculiar dynamic effects, neglecting sources of energy dissipation such as kinetic energy, viscous damping, and duration effects, (iv) difficulty in incorporating three-dimensional and cyclic earthquake loading effects.

In Figure 2 and Figure 3, examples of inadequate prediction of both the capacity curve as well as the deformation response characteristics of a *12*-storey reinforced concrete frame subjected to a natural earthquake recording (case-study RM15-NR2 in Antoniou and Pinho (2004a)) and of a *4*-storey irregular frame subjected to an artificial accelerogram (ICONS full-scale test specimen, described in Pinho and Elnashai (2000)) are given. It is noted that although the *12*-storey building is regular in height, its response is heavily influenced by higher mode effects, effectively rendering its seismic behaviour highly irregular in height, as conspicuously shown by Figure 2a. The standard pushover results have been carried out using both triangular and uniform loading distributions, and are compared with the envelope of results obtained with incremental dynamic analysis.

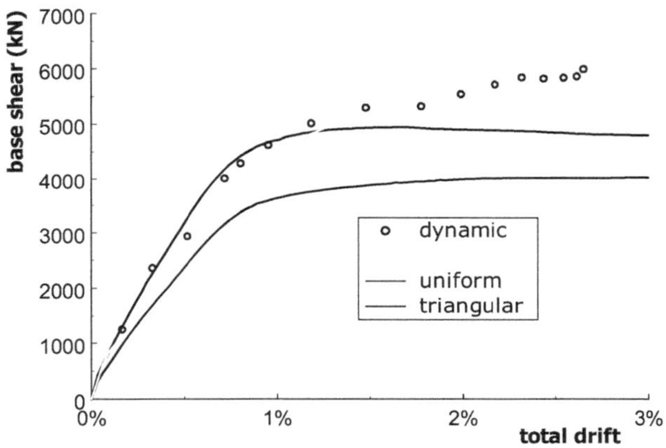

Figure 2. Capacity curves of a *12*-storey building, obtained with standard pushover.

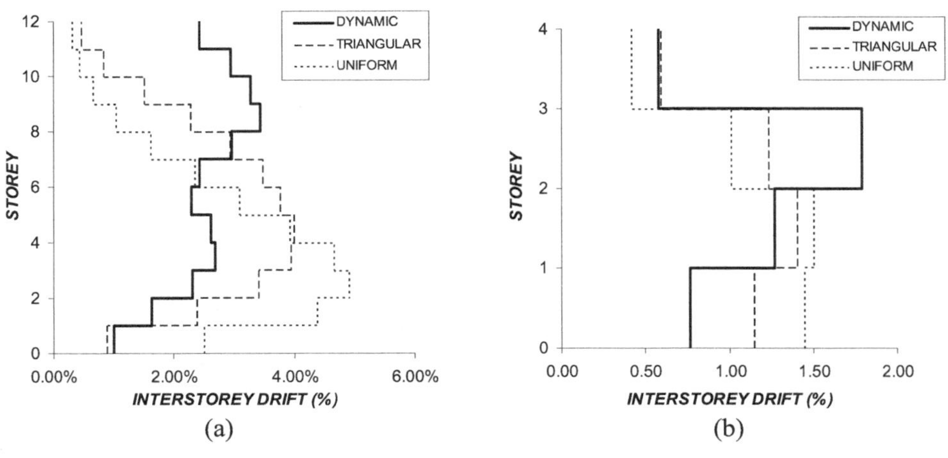

Figure 3. Interstorey drift profiles of (a) *12*-storey building and (b) *4*-storey irregular frame, obtained with standard pushover.

The main reason behind the underperformance of these conventional pushover methods is the fact that they do not account for the effect that damage accumulation, induced by the increasing deformation levels imposed on the structure, has on the response of the latter. Cumulative material straining introduces a reduction in stiffness, which, in turn, causes an elongation of the periods of vibration (Figure 4), which then, depending on the shape of the response spectrum being considered (or on the frequency content of an input record), may trigger significant changes in the response characteristics of the buildings (Figure 5). Krawinkler and Seneviratna (1998) summarised the above with a single statement: fixed load patterns in pushover analysis are limiting, be they first modal or multimodal derived, because no fixed distribution is able of representing the dynamic response throughout the full deformation range.

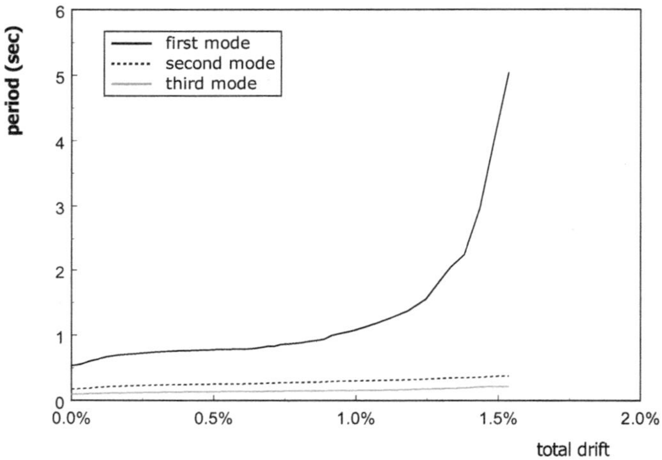

Figure 4. Periods of vibration of 4-storey building under increasing levels of deformation.

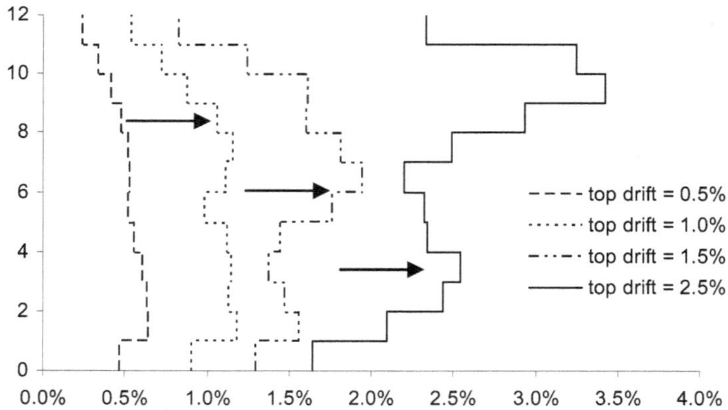

Figure 5. Interstorey drift profiles of a 12-storey building subjected to increasing levels of deformation.

As a result of the aforementioned limitations, recent years have witnessed the development and introduction of so-called Adaptive Pushover methods whereby the loading vector is updated at each analysis step, reflecting the progressive stiffness degradation of the structure induced by the penetration in the inelastic range (Figure 6). The response of the structure is thus computed in incremental fashion, through piecewise linearization (Figure 7), hence rendering it possible to use the tangent stiffness at the start of each increment, together with the mass of the system, to compute modal response characteristics of each incremental pseudo-system through elastic eigenvalue analysis, and use such modal quantities to congruently update (i.e. increment) the pushover loading vector.

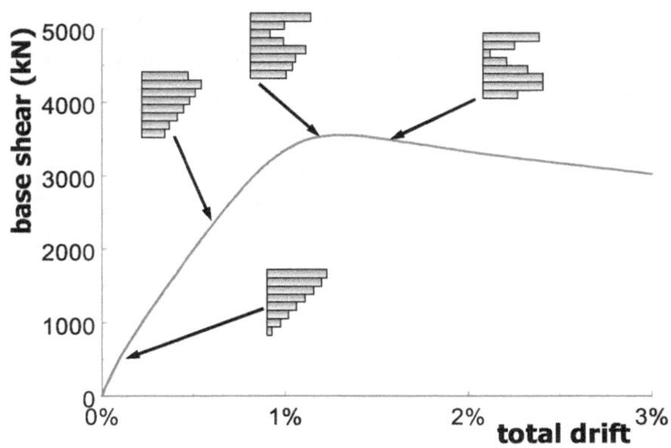

Figure 6. Adaptive pushover: shape of loading vector is updated at each analysis step.

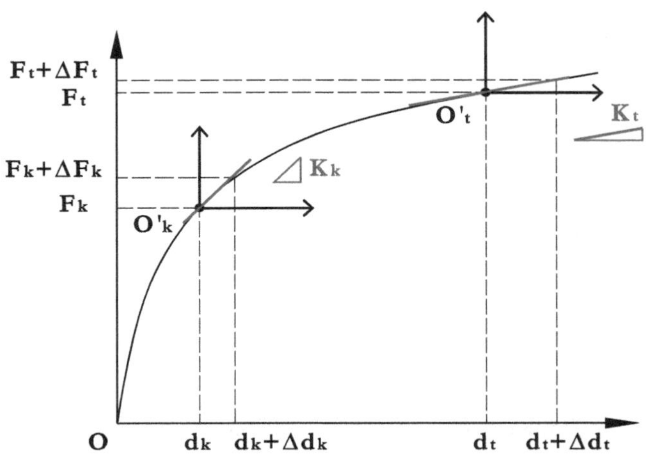

Figure 7. The use of tangent stiffness in updating (i.e. incrementing) the loading vector.

Force-based adaptive pushover procedures have been proposed by Reinhorn (1997), Bracci *et al.* (1997), Satyarno *et al.* (1998), Requena and Ayala (2000), Elnashai (2001) and Antoniou *et al.* (2002). With the exception of the work of Satyarno *et al.* (1998), where a single mode adaptive pushover pattern was employed, all other adaptive methodologies considered the effects of the higher modes and of the input frequency content. Furthermore, Elnashai (2001) and Antoniou *et al.* (2002) implemented their adaptive algorithm within a fibre analysis framework, allowing for a continuous, rather than discrete, force distribution update to be carried out. Despite their apparent conceptual superiority, or at least despite their conspicuously more elaborated formulation, the improvement introduced by such Force-based Adaptive Pushover (FAP) procedures was not-necessarily impressive, with respect to its traditional non-adaptive counterparts, particularly in what concerns the estimation of deformation patterns of buildings, which seemed to be poorly predicted by both types of analysis, as shown in Figure 8. As described by Kunnath (2004) and Antoniou and Pinho (2004a), the main reason for such underperformance seems to be the quadratic modal combination rules (e.g. SRSS, CQC) used in computing the adaptive updating of the load vector; such rules will inevitably lead to monotonically increasing load vectors, since the possibility of sign change in applied loads at any location is precluded, whilst it may be needed to represent the uneven redistribution of forces after an inelastic mechanism is triggered at some location.

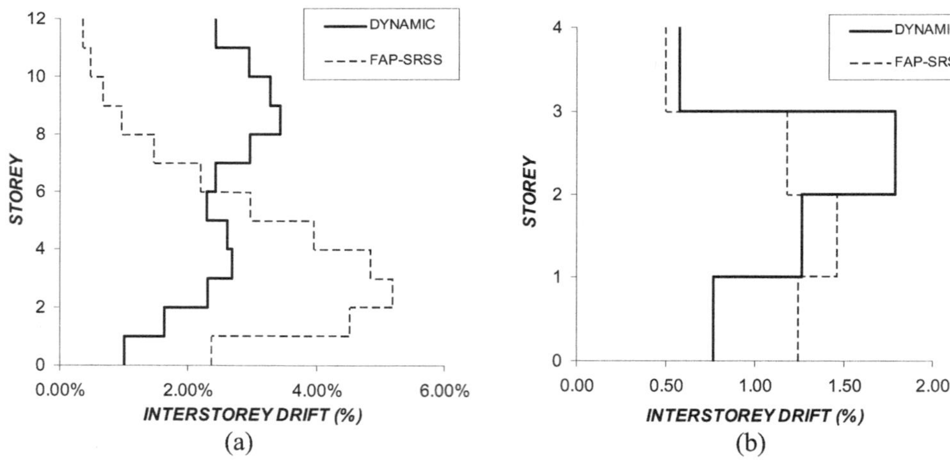

Figure 8. Interstorey drift profiles of (a) 12-storey building and (b) 4-storey irregular frame, obtained with Force-based Adaptive Pushover using SRSS modal combination.

With the above in mind, Kunnath (2004) and López-Menjivar (2004) have proposed an alternative modal combination scheme, consisting of a weighted Direct Vectorial Addition (DVA) of the different modal shapes that can be mathematically expressed as:

$$F_i = \sum_{j=1}^{n} \alpha_j \Gamma_j \phi_{j,i} M_j Sa_j \qquad \textbf{Eq. 1}$$

where i is the storey number, j is the mode number, n is the highest mode of interest, Γ_j is the modal participation factor for the jth mode, $\phi_{i,j}$ is the mass normalised mode shape value for the ith storey and the jth mode, M_i is the mass of the ith storey and Sa_j represents the acceleration response spectrum ordinate corresponding to the period of vibration of the jth mode. Finally, α_j is a weighting factor that aims at accounting for the varying relative importance that each mode j has on the maximum response of the structure.

The employment of such alternative modal combination procedure, may indeed lead to the attainment of improved results, as demonstrated by the interstorey drift profiles given in Figure 9, obtained through consideration of the first three modes of vibration of the buildings, and using $\alpha_1 = 1.0$, $\alpha_2 = -1.0$ and $\alpha_3 = 1.0$ in Eq. 1. However, the arbitrary nature of these weighting factors α_j renders the method unfeasible for practical application, as explicitly acknowledged in Kunnath (2004) and demonstrated in López-Menjivar (2004). Indeed, in the latter work it is demonstrated how values of α_j that lead to optimum results for some building configurations, lead then to poor predictions in buildings with diverse characteristics. Therefore, and until a general procedure to correctly determine the values of the weighting factors is found, the DVA adaptive pushover modality cannot really be deemed as a valid solution for practical application.

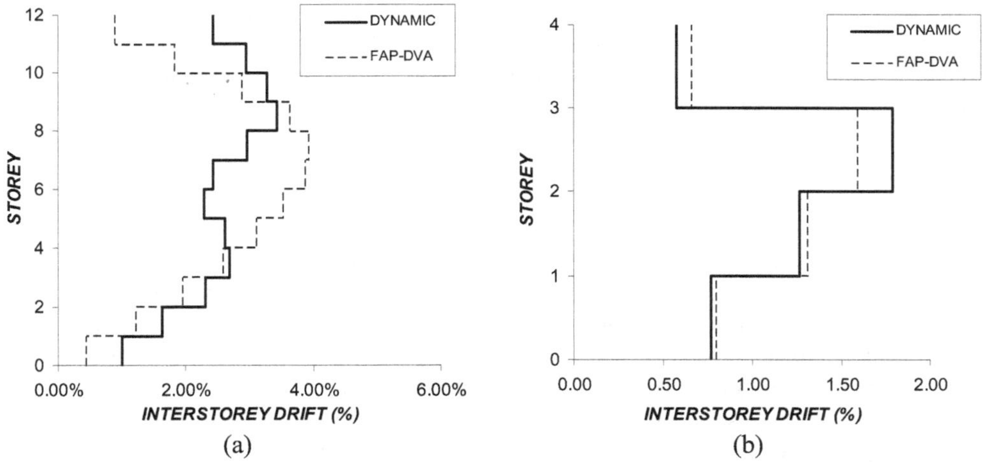

Figure 9. Interstorey drift profiles of (a) 12-storey building and (b) 4-storey irregular frame, obtained with Force-based Adaptive Pushover using DVA modal combination.

4 Displacement-Based Adaptive Pushover (DAP)

With a view to overcome all the limitations described above, Antoniou and Pinho (2004b) have proposed a paradigm shift in pushover analysis, by introducing the innovative concept of Displacement-based Adaptive Pushover (DAP). Contrarily to what happens in non-adaptive pushover, where the application of a constant displacement profile would force a predetermined and possibly inappropriate response mode, thus concealing important structural characteristics and concentrating inelastic mechanisms at a given location, within an adaptive framework, a displacement-based pushover is entirely feasible, since the loading vector is updated at each step of the analysis according to the current dynamic characteristics of the structure.

4.1 DAP – Methodology

The implementation of DAP can be structured in four main stages; (i) definition of the nominal load vector and inertia mass, (ii) computation of the load factor, (iii) calculation of the normalised scaling vector and (iv) updating of the loading displacement vector. Whilst the first step is carried out only once, at the start of the analysis, its three remaining counterparts are repeated at every equilibrium stage of the nonlinear static analysis procedure, as described in the following subsections.

The loading vector shape is automatically defined and updated by the solution algorithm at each analysis step, for which reason the nominal vector of displacements, U_0, must always feature a uniform (rectangular) distribution shape in height, so as not to distort the load vector configuration determined in correspondence to the dynamic response characteristics of the structure at each analysis step. In addition, it is noteworthy that the adaptive pushover requires the inertia mass M of the structure to be modelled, so that the eigenvalue analysis, employed in updating the load vector shape, may be carried out.

The magnitude of the load vector U at any given analysis step is given by the product of its nominal counterpart U_0, defined above, and the load factor λ at that step (see Eq. 2). The latter is automatically increased, by means of a load control strategy (Antoniou and Pinho, 2004a), until a predefined analysis target, or numerical failure, is reached.

$$U = \lambda \cdot U_0 \qquad \textbf{Eq. 2}$$

The normalized modal scaling vector, $\overline{D}$, used to determine the shape of the load vector (or load increment vector) at each step, is computed at the start of each load increment. In order for such scaling vector to reflect the actual stiffness state of the structure, as obtained at the end of the previous load increment, an eigenvalue analysis is carried out. To this end, the Lanczos algorithm (Hughes, 1987) is employed to determine the modal shape and participation factors of any given predefined number of modes. Modal loads can be combined by using either the Square Root of the Sum of the Squares (SRSS) or the Complete Quadratic Combination (CQC) methods.

Since application to the analysis of buildings is the scope of the present work, use is made of the interstorey drift-based scaling algorithm, whereby maximum interstorey drift values obtained directly from modal analysis, rather than from the difference between not-necessarily simultaneous maximum floor displacement values, are used to compute the scaling displacement vector. This comes as a reflection of the fact that the maximum displacement of a particular floor level, being essentially the relative displacement between that floor and the ground, provides insufficient insight into the actual level of damage incurred by buildings subject to earthquake loading. On the contrary, interstorey drifts, obtained as the difference between floor displacements at two consecutive levels, feature a much clearer and direct relationship to horizontal deformation demand on buildings. Readers are referred to Antoniou (2002) for further details on this formulation.

In such an interstorey drift-based scaling technique, the eigenvalue vectors are thus employed to determine the interstorey drifts for each mode Δ_{ij}, as shown in Eq. 3, while the displacement pattern D_i at the ith storey is obtained through the summation of the modal-combined inter-storey drifts of the storeys below that level, i.e. drifts Δ_1 to Δ_i:

$$D_i = \sum_{k=1}^{i} \Delta_k \quad \text{with} \quad \Delta_i = \sqrt{\sum_{j=1}^{n} \Delta_{ij}^2} = \sqrt{\sum_{j=1}^{n} \left[\Gamma_j \left(\phi_{i,j} - \phi_{i-1,j}\right)\right]^2} \qquad \textbf{Eq. 3}$$

Since only the relative values of storey displacements (D_i) are of interest in the determination of the normalised modal scaling vector $\overline{D}$, which defines the shape, not the magnitude, of the load or load increment vector, the displacements obtained by Eq. 3 are normalised so that the maximum displacement remains proportional to the load factor, as required within a load control framework:

$$\overline{D_i} = \frac{D_i}{\max D_i} \qquad \textbf{Eq. 4}$$

Once the normalised scaling vector and load factor have been determined, and knowing also the value of the initial nominal load vector, the loading vector U_t at a given analysis step t is obtained by adding to the load vector of the previous step, U_{t-1} (existing balanced loads), a newly derived load vector increment, computed as the product between the current load factor increment $\Delta\lambda_t$, the current modal scaling vector $\overline{D}_t$ and the nominal vector U_0, as mathematically translated into Eq. 5 and graphically depicted in Figure 10.

$$U_t = U_{t-1} + \Delta\lambda_t \cdot \overline{D_t} \cdot U_0 \qquad \textbf{Eq. 5}$$

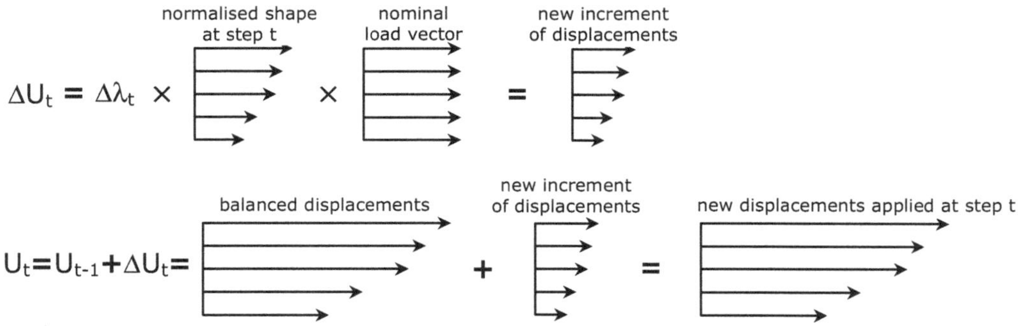

Figure 10. Updating of the loading displacement vector.

The DAP algorithm has been implemented in the computer code SeismoStruct (SeismoSoft 2005), a fibre element-based program for seismic analysis of framed structures, freely downloadable from the internet. The program incorporates both local (beam-column effects) and global (large displacements/rotations effects) sources of geometric nonlinearity as well as the interaction between axial force and transverse deformation of the element. The spread of material inelasticity along the member length is explicitly represented through the employment of a fibre modelling approach, implicit in the formulation of the inelastic beam-column frame elements adopted in the analyses. Various verification studies have been carried out with the aforementioned program on a four-storey reinforced concrete frame (Figure 11), a reinforced concrete bridge (Figure 12a) and a two-storey steel frame (Figure 12b) all of which show the ability of the analytical models to replicate the seismic response of full-scale structures.

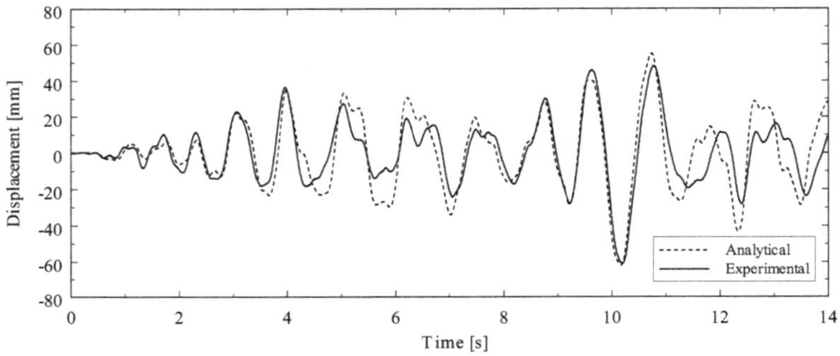

Figure 11. Verification of a fibre-element analytical model of a 4-storey RC frame (López-Menjivar, 2004).

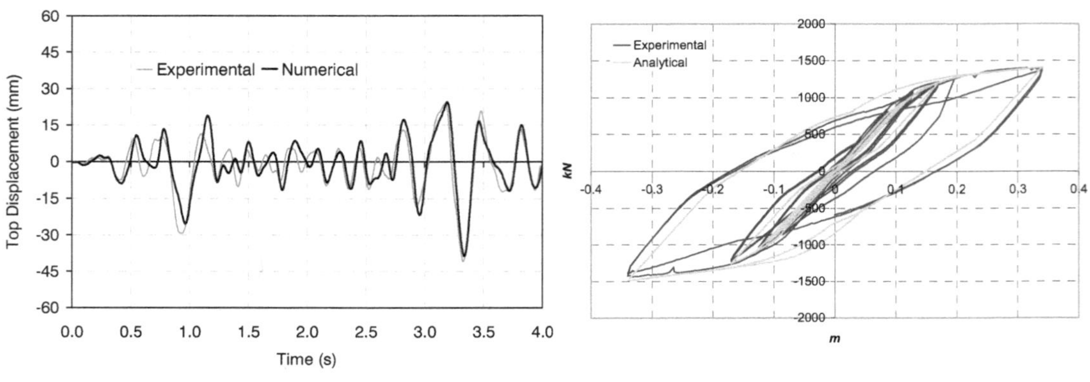

(a) (b)

Figure 12. Verification of fibre element analytical models of (a) a RC bridge (Casarotti *et al.*, 2005) and (b) a 2-storey steel frame (Pietra, 2006).

4.2 DAP – Illustrative results

Two clearly distinct building frames, both of which featuring an irregular type of dynamic response, are considered herein. The first of these is a *12*-storey five-bay reinforced concrete structure designed according to Eurocode 8 (2002). It displayed a highly irregular dynamic behaviour (Figure 3) when subjected to an accelerogram (Hollister station, Loma Prieta earthquake, USA, 1989) that presented a very high amplification in the short-period and thus lead to a response very much dominated by the *2nd* and *3rd* modes of vibration. Indeed, and as can be observed in Figure 13, these two higher modes (*0.15 < T2, T3 < 0.30 secs*) feature a spectral amplification, in acceleration, that is ten times higher than that corresponding to first mode of vibration (*T1 > 1.4 secs*). Further details on this case-study can be found in Antoniou *et al.* (2002).

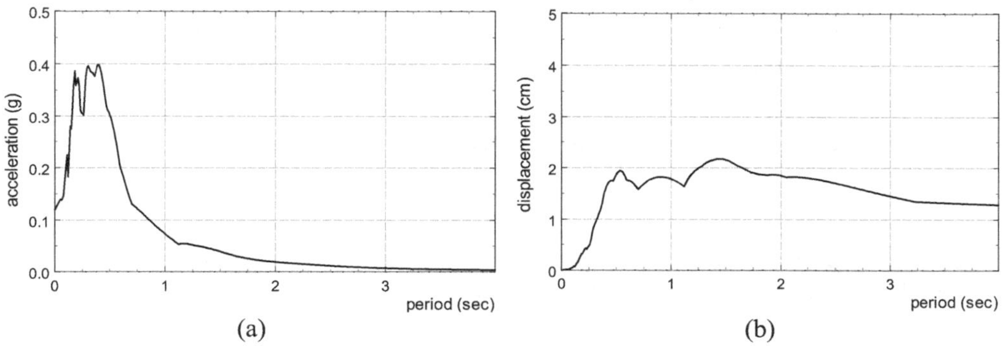

Figure 13. (a) Acceleration and (b) displacement response spectra of accelerogram employed in the analysis of 12-storey building.

The second structure is a *4*-storey three-bay building refers to a full-scale test specimen, built to represent typical design and construction practice in most South-European countries in the 1950's, and tested under pseudo-dynamic conditions (Pinho and Elnashai, 2000) at the JRC in Ispra (Italy). The frame was designed for gravity loads only, without any consideration of ductility provisions or capacity design principles. Consequently, it exhibited a soft-storey type of deformation mechanism at the third storey level (e.g. Figure 3) caused mainly by the drastic stiffness/strength variation present at such location, as well as by inadequate lap-splicing and defective column shear capacity. The input motion consisted of artificial accelerograms aiming at being representative of European seismicity (Campos-Costa and Pinto, 1999).

In Figure 14, the interstorey drift profiles of these two case-studies, as obtained with the employment DAP analyses, are given. It is observed that the predictions now match much closer the dynamic response of these two structures, which effectively means that the response irregularities caused by the flexibility of the *12*-storey structure, and subsequent amplification of higher modes, as well as the strength irregularity of the *4*-storey prototype, have been fully and correctly captured by the proposed static analysis algorithm.

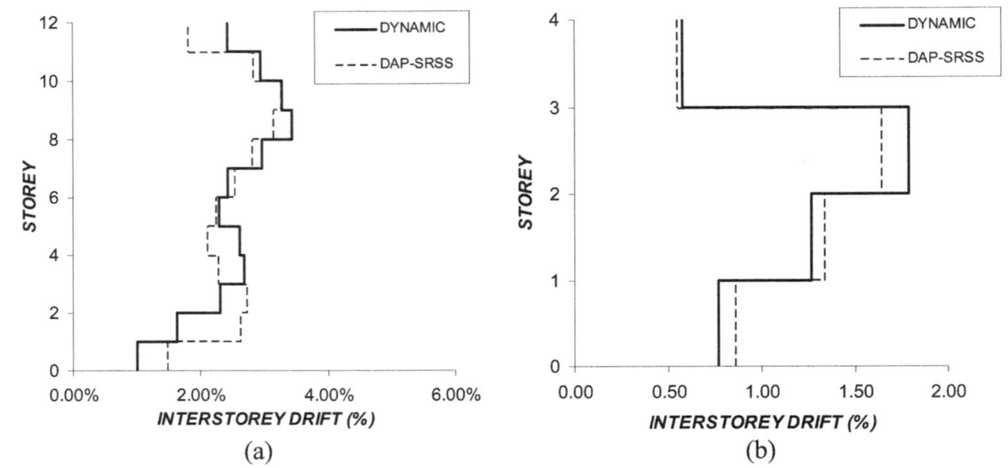

Figure 14. Interstorey drift profiles of (a) 12-storey building and (b) 4-storey irregular frame, obtained with Displacement-based Adaptive Pushover using SRSS combination.

In Figure 15, on the other hand, the capacity curves of the *12*-storey building, as derived by both DAP and standard pushover curves are compared with the Incremental Dynamic Analysis envelope. The advantages of using an adaptive displacement-based pushover can be inferred also from this type of results.

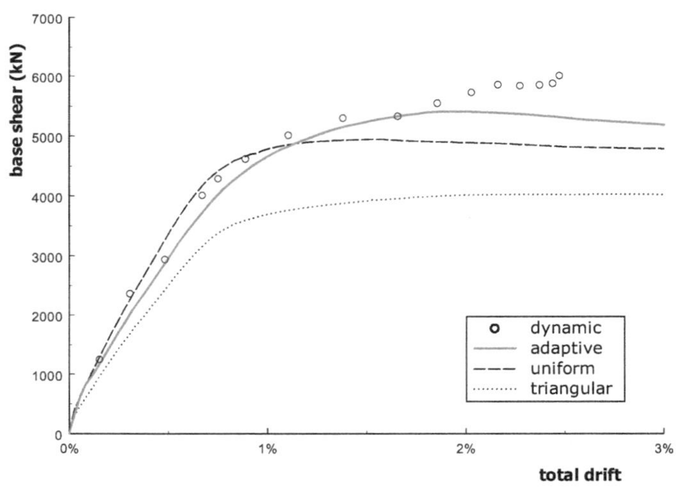

Figure 15. Capacity curves of a 12-storey building, obtained with DAP and standard pushovers, and compared against IDA envelopes.

The reason behind the most-improved predictions obtained with the displacement-based adaptive pushover procedure is the fact that storey forces or shears are no longer applied directly to the structure but rather come as a result of structural equilibrium to the applied displacement pattern, thus allowing for the reproduction of reversal of storey shear distributions, observed in dynamic analysis, even if a quadratic rule is employed to combine the contribution of the different modes. In effect, DAP drift profiles, despite carrying a permanently positive sign, do, in any case, feature changes of their respective gradient (i.e. the trend with which drift values change from one storey to the next), introduced by the contribution of higher modes. When such gradient variations imply a reduction of the drift of a given storey with respect to its adjacent floor levels, then the corresponding applied storey horizontal force must also be reduced, in some cases to the extent of sign inversion, as observed in Figure 16.

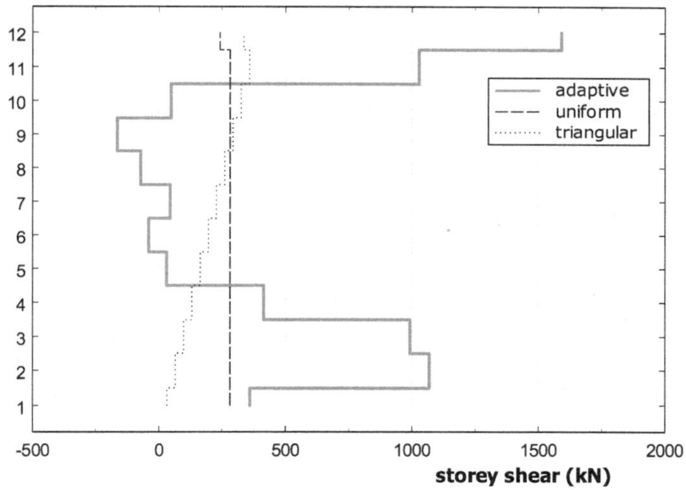

Figure 16. Storey shear distributions of a 12-storey building obtained with Displacement-based Adaptive Pushover as well as with standard non-adaptive pushovers.

In other words, given that in DAP, shear distributions are automatically derived to attain structural equilibrium with the imposed storey drifts, rather than being a result of the loads directly applied to the structure, the previously described limitations evidenced by force-based adaptive schemes that use quadratic modal combination rules can be overcome and, consequently, results as whole (i.e. deformation profiles and capacity curves) become more accurate.

4.3 DAP – Ease-of-use, Computational Effort and Numerical Stability

When compared with nonlinear time-history analysis, pushover methods are advantaged by their (i) higher user-friendliness, (ii) reduced running time and (iii) increased numerical stability. Therefore, it is important that the proposed displacement-based algorithm, capable of producing improved structural response predictions in comparison with existing non-adaptive pushover techniques, does also feature these three advantages over dynamic analysis.

From a usability point-of-view, the proposed displacement-based adaptive pushover algorithm effectively presents no additional effort and/or requirements with respect to its conventional non-adaptive counterparts. In effect, the only element of novelty, in terms of analysis input, is the introduction of the building's inertia mass, which, however, can readily be obtained directly from the vertical gravity loads, already included in any type of pushover analysis.

With regards to computational effort, in general, the amount of time necessary to complete an adaptive pushover analysis is typically double the time necessary for a conventional procedure, approximately. Obviously, the duration of such finite element runs will vary according to the computing capacity of the workstation being used, as well as with the characteristics of the model (mainly the number of elements and level of fibre discretisation of the sections). In any case, adaptive pushover proved to be up to ten times quicker than nonlinear dynamic analysis of a same model (keeping in mind that fibre-based finite element modelling has been adopted for the current work), hence the time-advantage of static methods versus their dynamic counterparts is not lost with the addition of the adaptive features.

As far as numerical stability is concerned, no particular problems have been observed in the studies described above, and those given in subsequent sections, noting that structures were pushed well into their post-peak inelastic response range.

5 Verification Parametric Studies

In this Section, an analytical comparative study involving different pushover methods, either single or multi mode, adaptive or conventional, and dynamic nonlinear analysis of buildings and bridges is presented. The "true" dynamic response is deemed to be represented by the results of the Incremental Dynamic Analysis procedure (IDA) (e.g. Vamvatsikos and Cornell, 2002), which is a parametric analysis method by which a structure is subjected to a series of nonlinear time-history analyses of increasing intensity.

5.1 Reinforced Concrete Frames

Characteristics of input motion and structural models. Three different configurations of common RC structures were employed: a *12*-storey regular frame, an eight storey irregular frame and a dual wall-frame system (Figure 17). The latter are based on buildings previously designed for different ductility classes and design ground acceleration, on medium soil type 'B' of EC8 (Fardis, 1994), resulting in a total of 12 models, as described in Table 1. The overall plan dimensions of the three configurations are *15 m* by *20 m*. The storey height is *3 m* except the first storey of the irregular set, which is *4.5 m* high. A detailed description of models and load conditions, as well as of their FE modelling, can be found in López-Menjivar (2004).

Four input time-histories, consisting of one-artificially generated accelerogram (A975) and three natural records (Loma Prieta earthquake, USA, 1989), were employed: the selection of these four records aimed at guaranteeing a wide-ranging type of earthquake action, in terms of frequency content, peak ground acceleration, duration and number of high amplitude cycles. Their acceleration response spectra are shown in Figure 18. Upper and lower bounds of the main characteristics of the records are summarised in Table 2, where the significant duration is defined as the interval between the build up of 5% and 95% of the total Arias Intensity (Bommer and Martinez-Pereira, 1999).

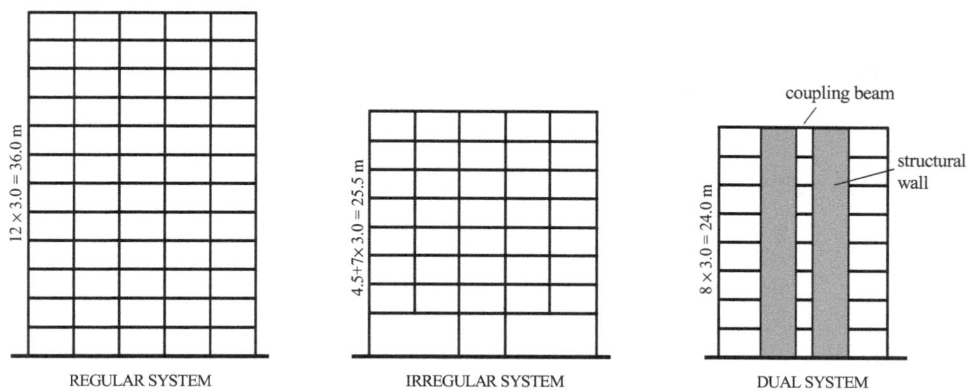

Figure 17. Geometric characteristics of the regular, irregular and dual systems.

Table 1. Considered building systems

Structural System	Storeys (Height)	Structure Reference	Ductility Level	Design PGA (g)	Behavior Factor (q)	$T_{uncracked}$ (s)
Regular Frame	12 (36 m)	RH30	High	0.30	5.00	0.697
		RM30	Medium		3.75	0.719
		RM15	Medium	0.15	3.75	0.745
		RL15	Low		2.50	0.740
Irregular Frame	8 (25.5 m)	IH30	High	0.30	4.00	0.565
		IM30	Medium		3.00	0.536
		IM15	Medium	0.15	3.00	0.613
		IL15	Low		2.00	0.614
Regular Wall-Frame	8 (24 m)	WH30	High	0.30	3.50	0.569
		WM30	Medium		2.625	0.557
		WM15	Medium	0.15	2.625	0.601
		RH30	High		5.00	0.697

Table 2. Bounding characteristics of the employed set of records for buildings

	Peak Ground Acceleration	Peak Response Acceleration	5% Arias Intensity threshold	Significant Duration t_{eff}	Total Duration t_{tot}	t_{eff} / t_{tot}
Min	0.12 g	0.50 g	1.02 s	7.24 s	10.0 s	22.3%
Max	0.93 g	4.25 g	11.23 s	10.43 s	40.0 s	72.4%

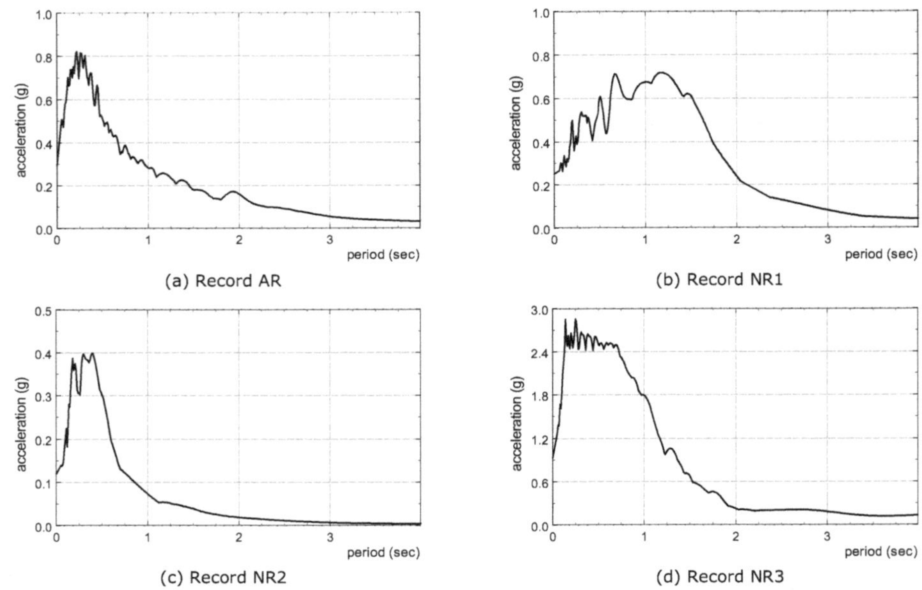

Figure 18. Elastic response spectra of the four records (5% equivalent viscous damping).

Analyses and results post-processing. The two non-adaptive pushover schemes, proposed in the NEHRP Guidelines (ATC, 1997), were applied to each set of buildings: the uniform distribution, whereby lateral forces are proportional to the total mass at each floor level, and the triangular distribution, in which seismic forces are proportional to the product of floor mass and storey height. The adaptive pushover algorithm was used in both its force and displacement-based variants, with spectrum scaling, employing SRSS or CQC modal combination rules.

The inter-storey drift profiles obtained from each pushover analysis are compared to the drift profiles from the nonlinear dynamic analysis and the standard error of the pushover results, with respect to the dynamic, is calculated as:

$$\mathrm{Error}(\%) = 100\sqrt{\frac{1}{n}\sum_{i=1}^{n}\left(\frac{\Delta_{iD} - \Delta_{iP}}{\Delta_{iD}}\right)} \qquad \textbf{Eq. 6}$$

The interstorey drift profiles are monitored at four different deformation levels: the pre-yield state (*0.5 %* total drift), the point of global yielding (*1.0 %* and *1.5 %*), where the stiffness changes significantly and the local distributions are rapidly updated, and the deeply inelastic range (*2.5 %*).

The Standard Error of the non-adaptive and adaptive pushover schemes was computed for all the structures and earthquake records considered. In order to identify the presence of possible response peculiarities introduced by individual input motions but smoothed out through results averaging, the standard error was computed separately for each time history analysis, as a unique value, averaging the standard error of all the storeys, in the building, and deformation levels.

Obtained results. The Mean Standard Error of the DAP, FAP, Triangular and Uniform pushovers, considering all structures and ground motions, are *19.11 %*, *30.90 %*, *21.11 %* and *38.76 %*, respectively. These overall results seem to indicate only a marginal advantage of DAP with respect to non-adaptive triangular distribution. However, a closer inspection of interstorey drift profiles (Figure 19) for some particularly difficult cases, renders much more conspicuous the gains provided by the employment of displacement-based adaptive pushover in the prediction of the seismic demand/capacity of framed buildings subjected to seismic action.

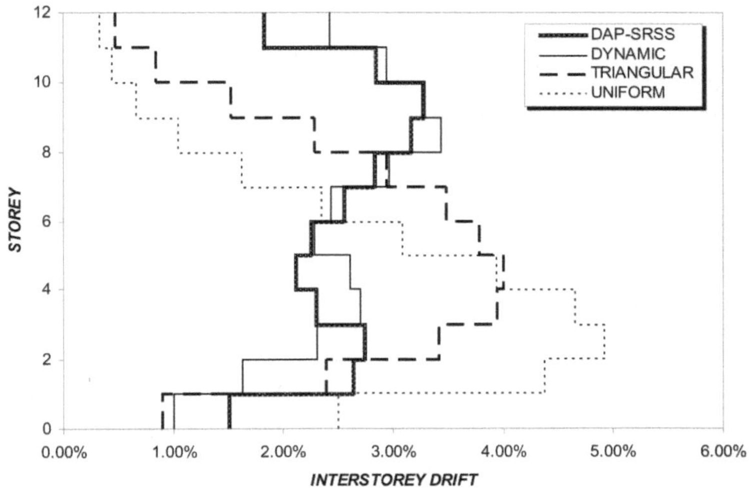

Figure 19. Representative results obtained with model RM15 subjected to one of the natural accelerograms employed in this study (NR2).

5.2 Steel Frames

Characteristics of input motion and structural models. In order to enable a direct comparison with the extensive and most thorough parametric study described in FEMA-440 (ATC, 2005), the same structural models and earthquake records that have been used in such work have been adopted here. Hence, considered ground motions include Near-Field type records (NF records) as well as ordinary records (Ordinary Ground Motions, OGMs). Their main properties are summarized in Table 3, whilst displacement response spectra are represented in Figure 20. OGMs (11 records) consist of site class C accelerograms selected from strong-motion records that do not present near-fault or near-field characteristics (strong velocity pulses, short duration, high frequency content, etc.). NF accelerograms, on the other hand, consist of motions recorded close to the epicentre and which contain very strong velocity pulses, originally included in the FEMA-440 work with the objective of verifying, at least in preliminary fashion, the validity of employing pushover methods in areas where pulse-like near-fault ground motions are likely to occur. For further details the reader is referred to ATC (2005) and Somerville *et al.* (1997).

Table 3. Ground Motion characteristics (ATC, 2005)

Identifier	EQ	M_s	Station	Component	PGA [g]	PGV [cm/s]	Effective Length[s]	Source
			Ordinary Ground Motions					
A1	Superstitn 11-24-87	6.6	El Centro Imp Co. Cent(01335)	000	0.358	46.4	23	CDMG
A2	Northridge 1-17-94	6.7	Canyon Country-W Lost Cany (90057)	000	0.41	43	11	USC
A3	Loma Prieta 10-18-89	7.1	Gilroy Array #2 (47380)	090	0.322	39.1	14	CDMG
A4	Chi-Chi, Taiwan 8-20-99	7.6	(TCU122)	N	0.261	34	35	CWB
A5	Loma Prieta 10-18-89	7.1	Gilroy Array #3 (47381)	090	0.367	44.7	17	CDMG
A6	Northridge 1-17-94	6.7	Canoga Park-Topanga Can (90053)	196	0.42	60.8	14	USC
A7	Chi-Chi, Taiwan 8-20-99	7.6	(CHY101)	W	0.353	70.6	32	CWB
A8	Superstitn 11-24-87	6.6	El Centro Imp Co. Cent(01335)	090	0.258	40.9	27	CDMG
A9	Northridge 1-17-94	6.7	Canoga Park-Topanga Can (90053)	106	0.356	32.1	16	USC
A10	Imperial Valley 10-15-79	6.9	El Centro Array #2 (5115)	140	0.315	31.5	17	USGS
A11	Imperial Valley 10-15-79	6.9	El Centro Array #11 (5058)	230	0.38	42.1	18	USGS
			Near-Field Ground Motions					
ERZ	Erzican 3-13-92	6.9	Erzican Station	NA	0.442	126	7	EERL Caltech
LUC	Landers 6-28-92	7.3	Lucerne Valley Station	280	0.732	147	14	EERL Caltech
RRS	Northridge 1-17-94	6.7	Rinaldi Receiving Station	213	0.891	186	6.5	EERL Caltech
SCH	Northridge 1-17-94	6.7	Sylmar County Hospital Parking Lot	190	0.865	138	5.5	EERL Caltech

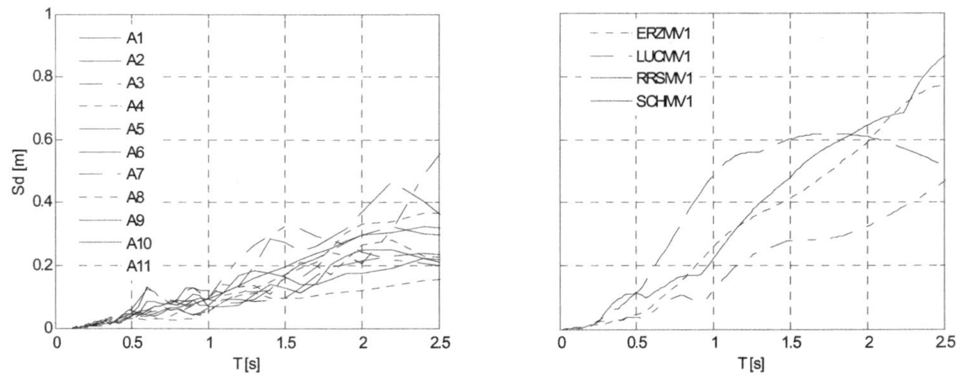

(a) (b)

Figure 20. Ground Motions: unscaled displacement response spectrum for (a) Ordinary- and (b) Near-Field-type records (ATC, 2005).

The prototype buildings analyzed in the current endeavour consist therefore of two steel moment-resisting frames (nine and three storeys, Figure 21 and Figure 22) designed as a part of the FEMA-funded SAC joint venture project (Gupta and Krawinkler, 1999). These frames were considered in both regular and irregular (weak-storey at the ground floor) configurations, thus leading to a total of four frames.

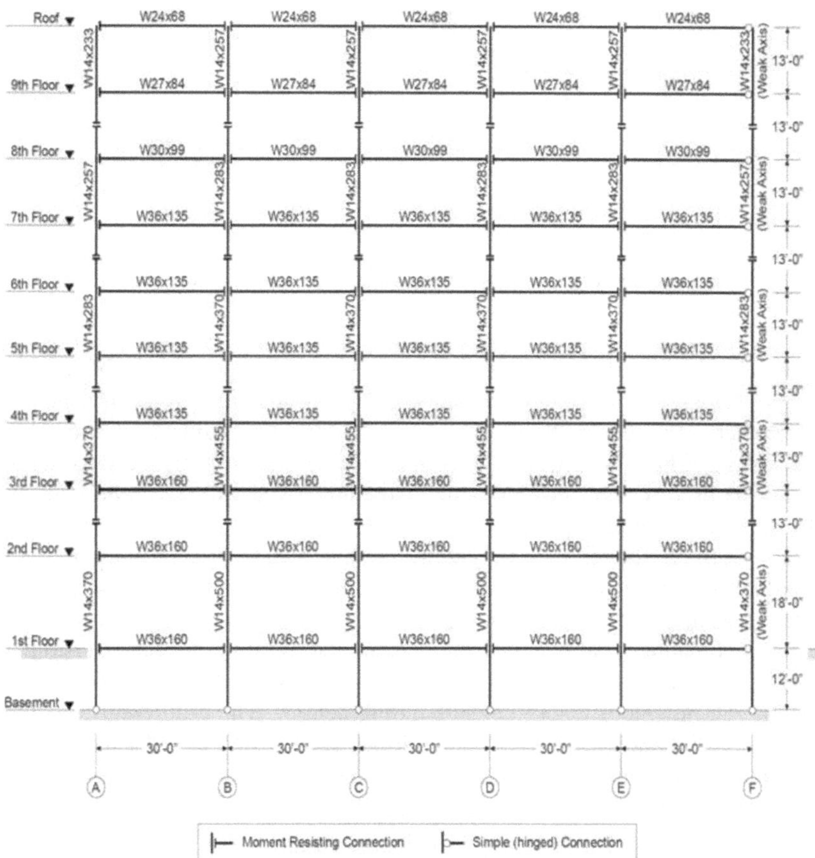

Figure 21. 9-Storey frame considered in this study (ATC, 2005).

Analyses and results post-processing. The pushover schemes considered include invariant static load patterns, such as (i) uniform [Uniform], (ii) inverted triangular [Triangular], (iii) first mode shape [1st Mode], (iv) a code-specified period-dependent distribution (where lateral forces changes from a linear distribution for low period systems to a parabolic shape for more flexible models) [Code], and the Displacement-based Adaptive Pushover [DAP]. In addition, an alternative adaptive approach has been tested in the case of OGMs, consisting in the employment of the

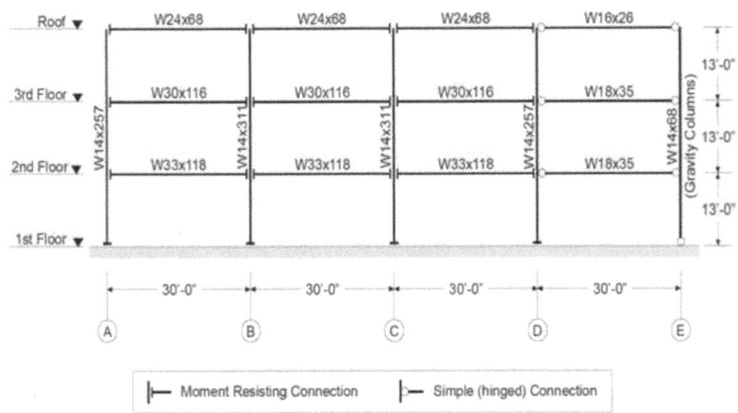

Figure 22. 3-Storey frame considered in this study (ATC, 2005).

average response spectrum of all records to compute the modal spectral amplification that is considered in the calculation of the incremental adaptive loading vector (Displacement-based Adaptive Pushover with Average Spectrum scaling [DAP-AS]). Whilst such procedure might be adopted in the case of OGMs, where records have been scaled to meet the same target displacement, it cannot be used for NF records, which were employed without scaling and thus involve different drift responses for the same prototype building.

Verification of the pushover algorithm at the "global level" was carried out through comparisons between the base shear vs. top floor IDA envelopes (assumed as representative of the true behaviour of the frame) and the pushover curves. With the objective of assessing also the accuracy of DAP in estimating local response parameters, the dynamic response at different ductility demand levels has been considered, where each Ductility Level (DL) is identified by means of the total drift value (i.e. top floor displacement/building height). Three different drift levels are assumed (0.5, 2 and 4%) and each ordinary record has been scaled in order to get the predefined drift for each prototype building. Response parameters of interest (displacement, drift, shear and moment) recorded in time-history analyses are then compared, at each level location, with those predicted by the pushover procedures at the same roof displacement magnitude. For the 9-Storey weak frame system an ultimate drift level of 2.7% has been selected (corresponding to a top floor displacement of 1m), instead of the 4% defined in a preliminary stage, since higher values were leading to the development of a global failure mechanism of the structure (corresponding to an ultimate steel strain in frame elements conservatively fixed in 15% (Ballio and Mazzolani 1987)) under several records, and would thus prevent a complete statistical evaluation of the results.

The effectiveness of the different static procedures in predicting the local dynamic response is quantified and compared by means of error measure E1, which provides a direct insight on how inaccurate is the static method (evidently, the mean of a pushover response estimate is computed only in the case of record-dependent DAP analysis, for all other pushover schemes the single response value is used):

$$E1 = \left| \frac{Mean_{PUSHOVER} - Mean_{DYNAMIC}}{Mean_{DYNAMIC}} \right| \quad \textbf{Eq. 7}$$

Obtained results. Figure 23 summarises the results obtained through this parametric study, putting in evidence the fact that, when compared with other pushover procedures, DAP leads to higher accuracy in the prediction of global and local response parameters of steel buildings, particularly in those cases where the influence of higher modes of vibration is important (e.g. high-rise buildings). It is also shown that the employment of an average response spectral shape leads to satisfactory results, thus rendering the procedure very much applicable within a design application framework, where standard code spectral shapes are prescribed.

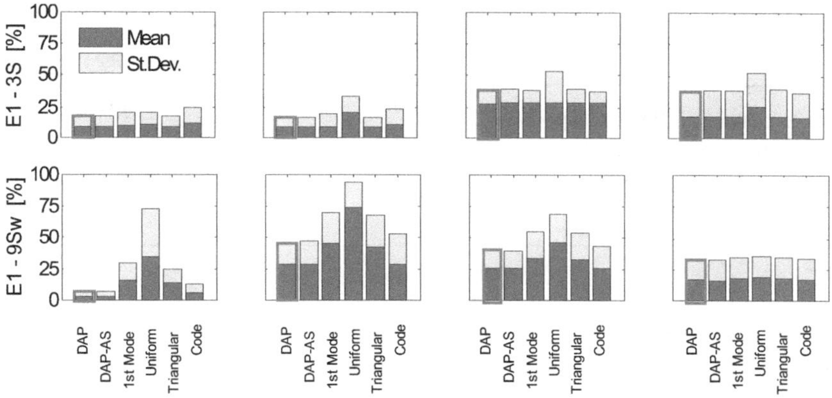

(a) Floor Displacement (b) Interstorey Drift (c) Interstorey Shear (d) Interstorey Moment

Figure 23. Mean values and standard deviation of error measure E1, averaged along the height of buildings models 3S and 9Sw at the intermediate drift level (top floor drift level of 2%).

5.3 RC Continuous-Span Bridges

Characteristics of input motion and structural models. The parametric study has considered two bridge lengths (*50 m* spans), with regular, irregular and semi-regular layout of the pier heights and with two types of abutments; (i) continuous deck-abutment connections supported on piles, exhibiting a bilinear behaviour, and (ii) deck extremities supported on pot bearings featuring a linear elastic response. The total number of bridges is therefore twelve, as shown in Figure 24, where the label numbers *1*, *2*, *3* characterise the pier heights of *7 m*, *14 m* and *21 m*, respectively.

A sufficiently large number of records has been employed so as to bound all possible structural responses. The employed set of seismic excitation is defined by an ensemble of *14* large magnitude (*6 ÷ 7.3*) small distance (*13 ÷ 30 km*) records selected from a suite of historical earthquakes scaled to match the *10 %* probability of exceedance in *50* years uniform hazard spectrum for Los Angeles (SAC Joint Venture, 1997). The bounding characteristics of the records are summarized in Table 4. Further details on modelling and input can be found in Casarotti *et al.* (2005).

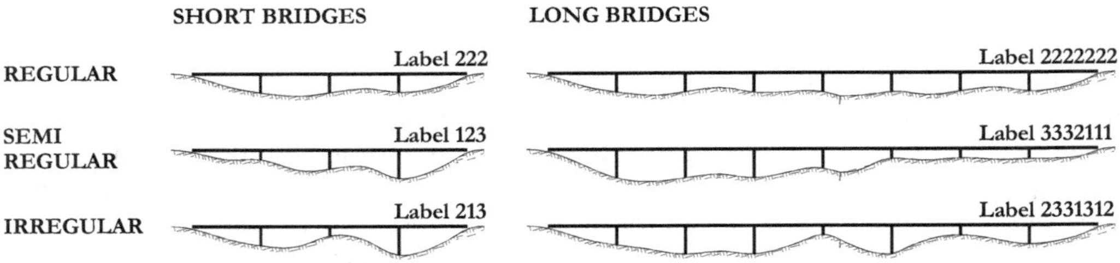

Figure 24. Analysed bridge configurations.

Table 4. Bounding characteristics of the employed set of records for bridges

	Peak Ground Acceleration	Peak Response Acceleration	5% Arias Intensity threshold	Significant Duration t_{eff}	Total Duration t_{tot}	t_{eff} / t_{tot}
Min	0.30 g	0.84 g	1.25 s	5.3 s	14.95 s	9%
Max	1.02 g	3.73 g	12.5 s	19.66 s	80.00 s	52%

Analyses and results post-processing. The response of the bridge models is estimated through the employment of Incremental Dynamic Analysis (IDA), Force-based Conventional Pushover with uniform load distribution (FCPu), Force-based Conventional Pushover with first mode proportional load pattern (FCPm), Force-based Adaptive Pushover with Spectrum Scaling (FAP) and Displacement-based Adaptive Pushover with Spectrum Scaling (DAP). Results are presented in terms of the bridge capacity curve, i.e. a plot of the reference point displacement versus total base shear, and of the deck drift profile.

Each level of inelasticity is represented by the deck centre drift, selected as independent damage parameter, and per each level of inelasticity the total base shear V_{base} and the displacements Δ_i at the other deck locations are monitored. Results of pushover analyses are compared to the IDA median value out of the responses to the *14* records, of each response quantity *R*, be it total base shear or deck drift:

$$\hat{R}_{i,IDA} = \text{median}_{j=1:14} \, R_{i,j-IDA} \Big] \qquad \textbf{Eq. 8}$$

Pushover analyses with spectrum scaling (i.e. adaptive pushovers) are statistically treated in an analogous way: medians of each response quantity represent that particular pushover analysis (i.e. FAP or DAP) with spectrum scaling. Finally, the results of each type of pushover are normalized with respect to the corresponding "exact" quantity obtained from the IDA medians, as schematically illustrated in Figure 25, and translated in Eq. 9. Representing results in terms of ratios between the "approximate" and the "exact" procedures, immediately indicates the bias in the approximate procedure, as the ideal target value of the different pushovers is always one.

$$\overline{R}_{i,PUSHOVERtype} = \frac{R_{i,PUSHOVERtype}}{\hat{R}_{i,IDA}} \cdots \xrightarrow[\text{ideally}]{} 1 \qquad \textbf{Eq. 9}$$

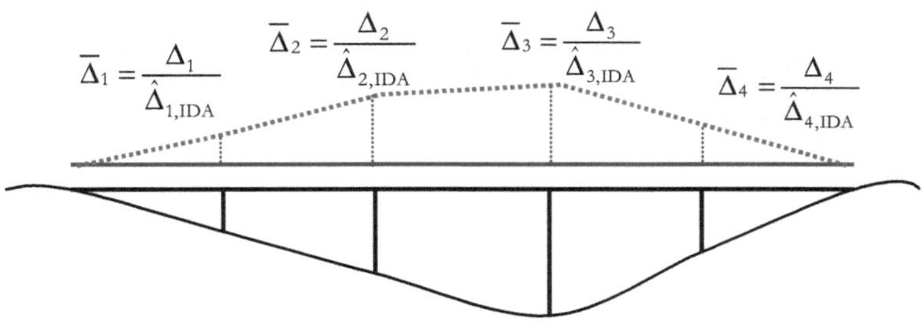

Figure 25. Normalised transverse deformed pattern.

Given the fact that a "realistic" capacity curve does not imply reliable estimations of the inelastic displacement pattern at increasing levels of inelasticity, the control of the deformed pattern is of the same relevance of the capacity curve prediction.

Having the same unitary target value, all normalized deck displacements become comparable, and a bridge index *BI* can measure the precision of the obtained deformed shape. Per each level of inelasticity, such bridge index is defined as the median of results over the *m* deck locations (Eq. 10), with the standard deviation measuring the dispersion with respect to the median (Eq. 11). The latter indicates the stability of the estimate of displacements along the deck: a small scatter means that predicted normalised displacements along the deck are averagely close to their median value *BI*.

$$BI_{PUSHOVERtype} = \text{median}_{i=1:m}(\overline{\hat{\Delta}}_{i,PUSHOVERtype}) \qquad \textbf{Eq. 10}$$

$$\delta_{PUSHOVERtype} = \sqrt{\frac{\sum_{i=1}^{m}\left(\overline{\hat{\Delta}}_{i,PUSHOVERtype} - BI_{PUSHOVERtype}\right)^2}{m-1}} \qquad \textbf{Eq. 11}$$

Obtained results. Current code recommendations require performing pushover analysis by pushing the entire structure with distributed load. In case of bridges, the additional option of pushing only the deck has been investigated, observing that the superstructure is the physical location where the most of the structural mass, i.e. the source of the inertia forces on the bridge, is usually concentrated and where it is relatively free to be excited. A preliminary investigation indicated a significant improvement in terms of stability and velocity of analysis in case of DAP and a very poor influence on results with the application of the latter option, which is thus recommended and employed in the parametric study.

Two main pertinent observations can be withdrawn from a scrutiny of the capacity curves obtained by the different pushover analyses in Figure 26: first, FCPm tends to significantly underestimate the structural stiffness, mainly due to the fact that, for the same base shear, central deck forces are generally higher compared to the other load patterns, thus results in larger displacement at that location. Then, on occasions, a "hardening effect" in the pushover curve occurs, which is sometimes reproduced only by employing DAP: once piers saturate their capacity, abutments absorb the additional seismic demand, proportionally increasing shear response and hardening the capacity curve.

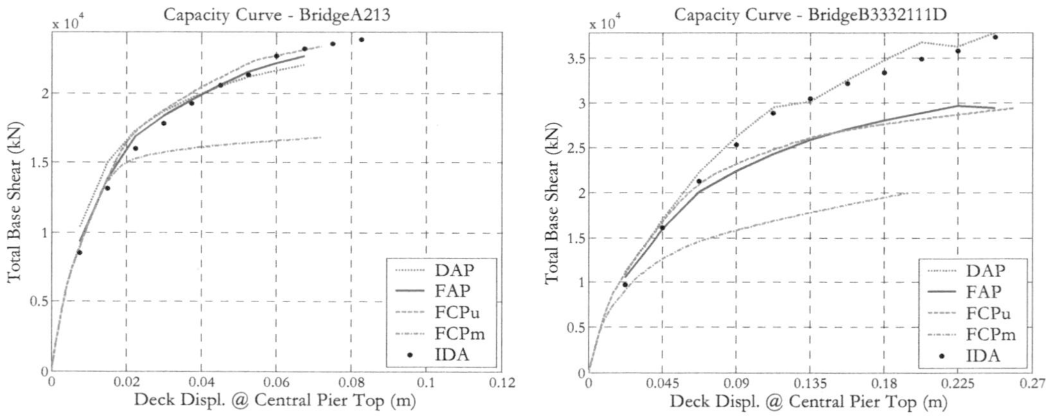

(a) (b)

Figure 26. Capacity curve results.

In Figure 27, the Bridge Index, as computed at each level of deck centre drift, is plotted as black filled marks so as to cater for an easier comparison with the IDA-normalised deck displacements, represented as empty marks in the background. In this manner, it results immediately apparent the level with which each pushover analysis is able to capture the deformed pattern of the whole bridge, at increasing deformation levels. For the sake of succinctness, only two analysis types are considered, FCPm and DAP, which are those leading to the worst and best predictions, respectively.

Table 5 provides global averages of means, maximum and minimum values of BI and respective dispersion as well as of the normalised total base shear, over the whole bridge ensemble. It is noted that: (i) FCPm heavily underestimates predictions, featuring also a very high BI dispersion value, (ii) FCPu performs very well for regular bridges and underestimating otherwise, (iii) DAP features the best overall behaviour, despite the slight underprediction of deformed shape values, with the lowest values of scatter.

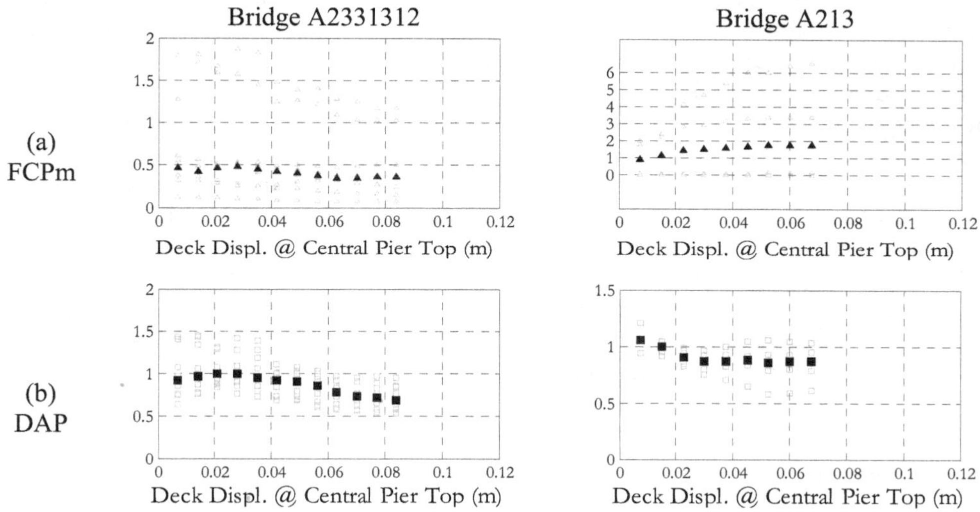

Figure 27. Prediction of the deformed pattern: BI and relative scatter.

Table 5. Global averages of the summaries of results

Means	Bridge Index			Dispersion			Normalised Base Shear		
	mean	min	max	mean	min	max	mean	min	max
FCPm	0.74	0.57	0.92	0.79	0.58	1.00	0.80	0.69	0.95
FCPu	0.87	0.75	1.03	0.24	0.17	0.34	1.03	0.92	1.18
FAP	0.88	0.78	1.01	0.22	0.13	0.34	0.99	0.89	1.10
DAP	0.87	0.78	0.99	0.19	0.14	0.27	1.03	0.95	1.13

6 Concluding Remarks

Given that current performance-based design trends require simple, yet accurate methods for estimating seismic demand on structures considering their full inelastic behaviour, in the current work the effectiveness of pushover analysis applied to buildings and bridges has been investigated. In particular, the effectiveness of applying a displacement-based adaptive pushover to estimate the seismic response of buildings and bridges subjected to earthquake action was investigated.

It is observed that the employment of such an innovative adaptive pushover technique lead to the attainment improved response predictions, throughout the entire deformation range, in comparison to those obtained by force-based methods, either adaptive or conventional. Indeed, prediction of the global behaviour (capacity curves), as well as of the deformed shapes and shear/moment distributions, proved to be very effective.

In other words, within the scope of buildings and bridge applications, whereas the application of a fixed displacement pattern is a commonly agreed conceptual fallacy, the present work witnesses not only the feasibility of applying an adaptive displacement profile, but also its practical advantages, with respect to other pushover methods.

It is important to observe that a static procedure will never be able to completely replace a dynamic analysis; nevertheless, a methodology has been searched to obtain response information reasonably close to that predicted by the nonlinear dynamic analyses. The innovative displacement-based adaptive pushover method is therefore shown to constitute an extremely appealing displacement-based tool for structural assessment, fully in line with the recently introduced deformation- and performance-oriented trends in the field of earthquake engineering.

Of equally noteworthy status is perhaps the fact that the proposed adaptive pushover scheme is as simple to use as standard pushover methods and has been implemented in an Internet-downloadable Finite Element program, thus giving the chance to researchers and practitioners to test and use it, where and when deemed opportune.

Acknowledgements

The author would like to acknowledge the important contribution of Dr C. Casarotti, Dr M. López-Menjivar and Mr D. Pietra for providing much of the material included in this manuscript. The author would also like to thank Dr S. Antoniou for the valuable assistance in the development and implementation of the adaptive pushover algorithm in the computer code employed in the current endeavour. Finally, the assistance of Mr D. Bolognini in the writing-up of this document is gratefully acknowledged.

References

Antoniou, S., (2002). Advanced inelastic static analysis for seismic assessment of structures, PhD Thesis, Imperial College London, University of London, United Kingdom.

Antoniou, S., Rovithakis, A., and Pinho, R., (2002). Development and verification of a fully adaptive pushover procedure, *Proc. 12th Europ. Conf. Earthq. Eng.*, London, UK, Paper No. 822.

Antoniou, S., and Pinho, R., (2004a). Advantages and limitations of adaptive and non-adaptive force-based pushover procedures, *J. Earthq. Eng.*, 8(4), 497-522.

Antoniou, S., and Pinho, R., (2004b). Development and verification of a displacement-based adaptive pushover procedure, *J. Earthq. Eng.*, 8(5), 643-661.

Applied Technology Council, (2005). *Improvement of Nonlinear Static Seismic Analysis Procedures*, FEMA-440, ATC, California, USA.

Applied Technology Council, (1997). *NEHRP Guidelines for the seismic rehabilitation of buildings*, FEMA Report No. 273, Federal Emergency Management Agency, Applied Technology Council, Washington D.C.

Aydinoglu, M. N., (2003). An incremental response spectrum analysis procedure based on inelastic spectral deformation for multi-mode seismic performance evaluation", *Bull. Earthq. Eng.*, 1(1), 3-36.

Ballio, G., and Mazzolani, F. M., 1987. *Strutture in acciaio*, Ulrico Hoepli editore S.p.A., Milano, Italy. (in Italian)

Bommer, J. J., and Acevedo, A. B., (2004). The use of real earthquake accelerograms as input to dynamic analysis, *J. Earthq. Eng.* 8 (Special Issue 1), 43-91.

Bommer, J. J., and Martinez-Pereira, A., (1999). The effective duration of earthquake ground motion, *J. Earthq. Eng.*, Vol. 3, Issue 2, pp.127-17

Bracci, J. M., Kunnath, S. K., and Reinhorn, A. M., (1997). Seismic performance and retrofit evaluation for reinforced concrete structures, *ASCE J. Struct. Eng.*, 123(1), 3-10.

Building Seismic Safety Council, (2000). *Prestandard and Commentary for the Seismic Rehabilitation of Buildings*, FEMA-356, Washington D.C., USA

Campos-Costa, A., and Pinto, A.V., (1999). *European Seismic Hazard Scenarios – An Approach to the Definition of Input Motions for Testing and Reliability Assessment of Civil Engineering Structures*, JRC Special Pubblication, no X.99.XX, ELSA, JRC – Ispra, Italy.

Casarotti, C., Pinho, R., and Calvi, G.M., (2005). *Adaptive Pushover-based Methods for Seismic Assessment and Design of Bridge Structures*, ROSE Report 2005/06, IUSS Press, Pavia, Italy.

Chopra A. K., and Goel R. K., (2002). A modal pushover analysis procedure for estimating seismic demands for buildings, *Earthq. Eng. Struct. Dyn.*, 31, 561-582.

Comité Europeen de Normalization, (2002). Eurocode 8: Design of Structures for Earthquake Resistance - Part 2: Bridges, PrEN 1998-2: 2003, 2 April 2002, CEN, Brussels, Belgium

Elnashai, A. S., (2001). Advanced inelastic static (pushover) analysis for earthquake applications, *Struct. Eng. Mech.*, 12(1), 51-69.

Fardis, M. N., (1994). Analysis and design of reinforced concrete buildings according to Eurocode 2 and 8. Configuration 3, 5 and 6, *Reports on Prenormative Research in Support of Eurocode 8.*

Goel, R. K., and Chopra, A. K., (2005a). Response to B. Maison's Discussion of "Evaluation of Modal and FEMA Pushover Analysis: SAC Buildings", *Earthq. Spectra*, 21(1), 277-279.

Goel, R. K., and Chopra, A. K., (2005b). Role of Higher-"Mode" Pushover Analyses in Seismic Analysis of Buildings, *Earthq. Spectra* 21(4), 1027-1041.

Gupta, B., and Kunnath, S. K., (2000). Adaptive spectra-based pushover procedure for seismic evaluation of structures, *Earthq. Spectra*, 16(2), 367-392.

Gupta, A., and Krawinkler, H., (1999). *Seismic demands for performance evaluation of steel moment resisting frame structures (SAC Task 5.4.3)*, Report No. 132, John A. Blume Earthquake Engineering Center, Stanford University, Stanford, California.

Hernández-Montes E., Kwon, O. S., and Aschheim, M., (2004). An energy-based formulation for first and multiple-mode nonlinear static (pushover) analyses, *J. Earthq. Eng.*, 8(1), 69-88.

Hughes, T. J. R., (1987). *The Finite Element Method, Linear Static and Dynamic Finite Element Analysis*, Prentice-Hall, Inc.

Krawinkler H., and Seneviratna G. D. P. K., (1998). Pros and cons of a pushover analysis of seismic performance evaluation, *Eng. Struct.*, 20(4-6), 452-64.

Kunnath, S. K., (2004). Identification of modal combination for nonlinear static analysis of building structures, *Computer-Aided Civil and Infrastruc. Eng.*, 19, 246-259.

Kunnath, S. K., and John, A. Jr., (2000). Validity of static procedures in performance-based seismic design, *Proceedings of ASCE Structures Congress*, Philadelphia, USA.

López-Menjivar, M. A., (2004). Verification of a displacement-based Adaptive Pushover method for assessment of 2-D Reinforced Concrete Buildings, PhD Thesis, European School for Advances Studies in Reduction of Seismic Risk (ROSE School), University of Pavia, Italy.

Maison, B., (2005). Discussion of "Evaluation of Modal and FEMA Pushover Analysis: SAC Buildings", *Earthq. Spectra* 21(1), 275-275.

Matsumori, T., Otani, S., Shiohara, H., and Kabeyasawa, T., (1999). Earthquake member deformation demands in reinforced concrete frame structures, *Proc., US-Japan Workshop PBEE Methodology for R/C Building Structures*, PEER Center Report, UC Berkeley - 79-94, Maui, Hawaii.

Moghadam, A. S., and Tso, W. K, (2002). A pushover procedure for tall buildings, Proc., *12th Europ. Conf. Earthq. Eng.*, London, UK, Paper 395.

Reinhorn, A., (1997). Inelastic analysis techniques in seismic evaluations, in *Seismic design methodologies for the next generation of codes*, Krawinkler and Fajfar (editors), Balkema, 277-287.

Paret, T. F., Sasaki, K. K., Eilbeck, D. H., and Freeman, S. A., (1996). Approximate inelastic procedures to identify failure mechanisms from higher mode effects, *Proc., 11th World Conf. Earthq. Eng.*, Acapulco, Mexico, Paper 966.

Pietra, D., (2006). *Evaluation of pushover procedures for the seismic design of buildings*, MSc Dissertation, European School for Advanced Studies in Reduction of Seismic Risk (ROSE School), University of Pavia, Pavia, Italy.

Pinho, R., and Elnashai, A. S., (2000). Dynamic collapse testing of a full-scale four storey RC frame, *ISET J. Earthq. Eng., Special Issue on Experimental Techniques*, 37(4), 143-164.

Requena, M., and Ayala, G., (2000). Evaluation of a simplified method for the determination of the nonlinear seismic response of RC frames, *Proc. 12th World Conf. Earthq. Eng.*, Auckland, New Zealand, Paper No. 2109.

SAC Joint Venture, (1997). *Develop Suites of Time Histories*, Project Task: 5.4.1, Draft Report, March 21, 1997, Sacramento, CA, USA

Sasaki, K. K., Freeman S. A., and Paret T. F., (1998). Multi-mode pushover procedure (MMP) – a method to identify the effects of higher modes in a pushover analysis, Proc., *6th US Conf. Earthq. Eng.*, Seattle, Washington – Earthquake Engineering Research Institute, Oakland, California.

Satyarno, I., Carr, A. J., and Restrepo, J., (1998). Refined pushover analysis for the assessment of older reinforced concrete buildings, *Proc. NZSEE Technology Conference*, Wairakei, New Zealand, 75-82.

Seismosoft, (2005). SeismoStruct - A Computer Program for Static and Dynamic nonlinear analysis of framed structures [online], available from URL: http//www.seismosoft.com.

Sommerville P., Smith N., Punyamurthula S. and Sun J., (1997). *Development of ground motion time histories for phase 2 of the FEMA/SAC steel project*, SAC Background Document, Report No. SAC/BD/-7/04, Applied Technology Council, Redwood City, CA

Vamvatsikos D., and Cornell C. A., (2002). Incremental dynamic analysis, *Earthq. Eng. Struct. Dyn.*, 31(3), 491-514.

The Need for Displacement-Based Design and Analysis *

Nigel Priestley[1]

[1] Emeritus Professor of Structural Engineering, University of California, San Diego

Abstract. A brief description of current force-based design is given. The historical basis for current seismic design philosophy is mentioned. Conceptual problems for force-based design using initial stiffness to represent structural response, which are often not recognized by designers are discussed.

1 Equivalence of Force and Displacement in Elastic Design

Although it has been traditional to consider seismic design in terms of forces, for reasons that will be discussed shortly, it can easily be shown that for elastic systems, it is at least as logical to design with a starting point of displacement. Consider the equivalent acceleration and displacement design spectra for 5% damping presented in Figure 1.

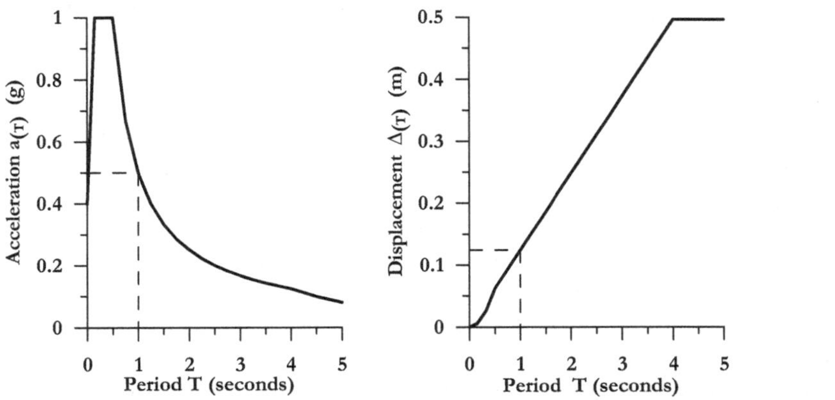

(a) Acceleration Spectrum for 5% damping (b Displacement Spectrum for 5% damping

Figure 1. Equivalent Acceleration and Displacement Response Spectrum.

Traditional seismic design has been based on the elastic acceleration spectrum (Figure 1(a). For an elastically responding single-degree-of-freedom (SDOF) structure, the response acceleration, a(T), corresponding to the fundamental period T, is found and the corresponding force, F and displacement Δ are given by

$$F = m \ a_{(T)} \ g; \ \ \Delta = F / K \tag{1}$$

where K is the system stiffness, m is the system mass and g is the acceleration due to gravity.

Alternatively the displacement spectrum of Figure 1(b) could be used directly. In this case the response displacement $\Delta(T)$ corresponding to the elastic period is directly read, and the corresponding force calculated as

$$F = K \ \Delta_{(T)} \tag{2}$$

In both cases the elastic period must first be calculated, but it is seen that working from the displacement spectrum requires one less step of calculation than working from the acceleration spectrum, since the mass is not needed once the period has been calculated. Although both approaches are directly equivalent, it would seem that using response displacement rather than response acceleration would be a more logical basis for design of elastic systems. We will show later that the reasons for using displacement spectra become more compelling when the seismic response of inelastic systems is considered.

2 Historical Considerations

The reason that seismic design is currently based on force (and hence acceleration) rather than displacement, is, as discussed above, based largely on historical considerations. Prior to the 1930's, few structures were specifically designed for seismic actions. In the 1920's and early 1930's several major earthquakes occurred (Japan: 1925 Kanto earthquake, USA: 1933 Long Beach earthquake, New Zealand: 1932 Napier earthquake). It was noted that structures that had been designed for lateral wind forces performed better in these earthquakes than those without specified lateral force design. As a consequence, design codes started to specify that structures in seismic regions be designed for lateral inertia forces. Typically, a value of about 10% of the building weight, regardless of building period, applied as a vertically distributed lateral force vector, proportional to the mass vector, was specified.

During the 1940's and 1950's, the significance of structural dynamic characteristics became better understood, leading to period-dependent design lateral force levels in most seismic design codes, by the 1960's. Also in the 1960's with increased understanding of seismic response, and the development of inelastic time-history analysis, came awareness that many structures had survived earthquakes capable of inducing inertia forces many times larger than those corresponding to the structural strength. This lead to development of the concept of ductility to reconcile the apparent anomaly of survival with apparently inadequate strength. Relationships between ductility and force-reduction factor, such as the "equal displacement" approximation (generally considered applicable for medium-period and long-period structures, and the "equal energy" approximation, which appeared more appropriate for short-period structures, were developed as a basis for determining the appropriate design lateral force levels.

During the 1970's and 1980's much research effort was directed to determining the available ductility capacity of different structural systems. Ductility considerations became a fundamental part of design, and key text books written in the 1960's and 1970's [e.g. Newmark and Rosenbleuth,1966; Park and Paulay,1976] have remained as the philosophical basis for seismic design, essentially till the present time. In order to quantify the available ductility capacity, extensive experimental and analytical studies were performed to determine the safe maximum displacement of different structural systems under cyclically applied displacements. This may be seen as the first turning away from force as the basis for design. Required strength was determined from a force-reduction factor that reflected the perceived ductility capacity of the structural system and material chosen for the design. Nevertheless, the design process was still carried out in terms of required strength, and displacement capacity, if directly checked at all, was the final stage of the design. The concept of "capacity design" was introduced (Park and Paulay,1976), where locations of preferred flexural plastic hinging were identified, and alternative undesirable locations of plastic hinges, and undesirable modes of inelastic deformation, such as shear failure, were inhibited by setting their strength higher than the force levels corresponding to that of the desired inelastic mechanism. Ductility was perceived as more important than displacement capacity, though the two were clearly related.

In the 1990's, textbooks with further emphasis on displacement considerations and capacity design became widely used for seismic design of concrete and masonry structures [e.g. Paulay and Priestley,1992; Priestley et al, 1996], and the concept of "performance-based seismic design", based largely on displacement considerations, became the subject of intense research attention. It may be seen from this brief description of the history of seismic design, that initially design was purely based on strength, or force considerations using assumed rather than valid estimates of elastic stiffness. As the importance of displacement has come to be better appreciated in recent years, the approach has been to attempt to modify the existing force-based approach to include consideration of displacement, rather than to rework the procedure to be based on a more rational displacement basis. These lectures present a different viewpoint, based on a complete re-examination of the fundamental basis of seismic design.

3 A Brief Review of Force-Based Seismic Design

Although current force-based design is considerably improved compared with procedures used in earlier years, there are many fundamental problems with the procedure, particularly when applied to reinforced concrete or reinforced masonry structures. In order to examine these problems, it is first necessary to briefly review the force-based design procedure, as currently applied in modern seismic design codes.

The sequence of operations required in force-based seismic design can be summarized as follows:

1. The structural geometry, including member sizes is estimated. In many cases these may be dictated by non-seismic load considerations.
2. Member elastic stiffnesses are estimated, based on preliminary estimates of member size. Different assumptions are made in different seismic design codes about the appropriate stiffnesses for reinforced concrete and masonry members. In some cases gross (uncracked section) stiffnesses are used, while in some codes reduced section stiffnesses are taken, to reflect the softening caused by expected cracking when approaching yield-level response.

3. Based on the assumed member stiffnesses, the fundamental period (equivalent lateral force approach) or periods (multi-mode dynamic analysis) are calculated. For a SDOF representation of the structure, the fundamental period is given by:

$$T = 2\pi\sqrt{\frac{m_e}{K}} \quad \textbf{(3)}$$

where m_e is the effective seismic mass (normally taken as the total mass).

In some building codes a height-dependent fundamental period is specified, independent of member stiffness, mass distribution, or structural geometry. The typical form of this is given in equation (4):

$$T = C_1(H_n)^{0.75} \quad \textbf{(4)}$$

where C_1 depends on the structural system, and H_n is the building height.

Lateral force levels calculated from stiffness-based periods (single mode or multi-mode) are not permitted to deviate from the forces based on the height-dependent period equation by more than some specified percentage.

4. The design base shear VBE for the structure corresponding to elastic response with no allowance for ductility is given by an equation of the form

$$V_{BE} = C_T \ I \ (gm_e) \quad \textbf{(5)}$$

where C_T is the basic seismic coefficient dependent on seismic intensity, soil conditions and period T (e.g. Figure 1(a)), I is an importance factor reflecting different levels of acceptable risk for different structures, and g is the acceleration of gravity.

5. The appropriate force-reduction factor R_μ (known in Europe as the behaviour factor Q) corresponding to the assessed ductility capacity of the structural system and material is selected. Generally R_μ is specified by the design code and is not a design choice, though the designer may elect to use a lesser value than the code specified one.
6. The design base shear force is then found from

$$V_{BR} = \frac{V_{BE}}{R_\mu} \quad \textbf{(6)}$$

The base shear force is then distributed to different parts of the structure to provide the vector of applied seismic forces. For building structures, the distribution is typically proportional to the product of the height and mass at different levels, which is compatible with the displaced shape of the preferred inelastic mechanism (beam-end plastic hinges plus column-base plastic hinges for frames; wall-base plastic hinges for wall structures). The total seismic force is distributed between different lateral force-resisting elements, such as frames and structural walls, in proportion to their elastic stiffness.

7. The structure is then analyzed under the vector of lateral seismic design forces, and the required moment capacities at potential locations of inelastic action (plastic hinges) is determined. The final design values will depend on the member stiffnesses.
8. Structural design of the member sections at plastic hinge locations is carried out, and the displacements under the seismic action are estimated.
9. The displacements are compared with code-specified displacement limits.
10. If the calculated displacements exceed the code limits, redesign is required. This is normally effected by increasing member sizes, to increase member stiffness.
11. If the displacements are satisfactory, the final step of the design is to determine the required strength of actions and members that are not subject to plastic hinging. The process known as capacity design [see Paulay and Priestley,1996] ensures that the dependable strength in shear, and the moment capacity of sections where plastic hinging must not occur, exceed the maximum possible input corresponding to maximum feasible strength of the potential plastic hinges. Most codes include a prescriptive simplified capacity design approach.

The above description is a simplified representation of current force-based design. In many cases the force levels are determined by multi-modal analysis (sometimes called dynamic analysis). The way in which the modal contributions are combined will be discussed in some detail in sections relating to different structural systems. Some design codes, such as the New Zealand Loadings Code define inelastic acceleration design spectra that directly include the influence of ductility rather than using an elastic spectrum and a force-reduction factor.

4 Inherent Problems with Force-based Seismic Design

4.1 Initial Stiffness Estimation

In force-based seismic design, the initial stiffness is used to estimate the period(s) of the structure, and hence the seismic forces. It is also used to distribute the design seismic force between structural elements. It is generally assumed in seismic design that if the dimensions of a structure are known, the stiffness can be directly estimated with sufficient accuracy for these purposes. That is, stiffness is independent of strength, if the dimensions are known. Required global strength is thus a function of the assumed global stiffness, and required local strength depends on the assumption of relative local stiffness.

When concrete or masonry structures are considered, it is found that the assumption of stiffness being independent of strength is invalid. Adopting the usual assumption that initial stiffness should be estimated at first-yield of reinforcement, it is found [Priestley,2003] that the yield curvature of a bilinear approximation to the moment-curvature response of a given section is essentially constant over a wide range of reinforcement and axial load levels. The stiffness of the section is thus proportional to strength, and may vary over a wide range. An example is presented in Fig.2 which plots stiffness of a rectangular column as a fraction of the gross-section stiffness (ignoring the stiffening effect of flexural reinforcement), as influenced by flexural reinforcement ratio, ρ_l, and axial load ratio N_u/f'_cA_g. A_s is seen from this example, the range of stiffness values corresponding to realistic reinforcement and axial load limits is eight-fold. To assume that stiffness is independent of strength is clearly an unacceptably coarse assumption.

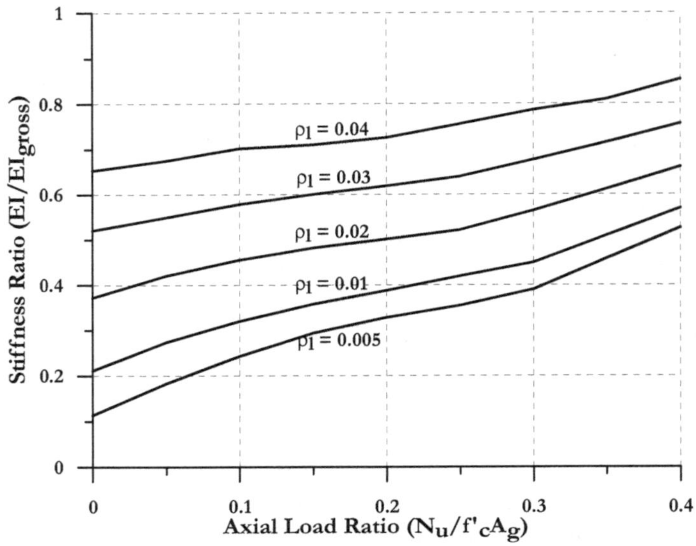

Figure 2 Effective Stiffness Ratio for Large Rectangular Reinforced Concrete Columns.

Analyses have shown that the yield curvature for a given reinforced concrete or masonry section can be expressed in the form:

$$\phi_y = C \ \varepsilon_y / h \tag{7}$$

where C is a constant dependent on the section shape, ε_y is the yield strain of the flexural reinforcement and h is the gross section depth [Priestley,2003]. The constant C varies little between different section shapes and is generally in the range $1.8 < C < 2.3$. It has also been found that the yield drift of concrete frames is essentially independent of strength [Priestley,1998], and can be expressed by the rather simple expression

$$\theta_y = 0.5\varepsilon_y \frac{l_b}{h_b} \tag{8}$$

where l_b and h_b are characteristic beam length and depth respectively.

Since structural stiffness is proportional to strength, conventional design allocating strength on the basis on an assumed stiffness is inappropriate. It is clear that design can be improved by an iterative process, where the stiffness is modified after a preliminary design has been undertaken, and a new estimate of the strength based on this stiffness is determined. This is time-consuming, and in our experience is rarely adopted.

4.2 Use of a Constant Behaviour Factor

Specification of a constant behaviour, or force reduction, factor for a given structural type and material (e.g. concrete structural wall), as is common in most seismic design codes does not account for the fact that ductility capacity is also a function of geometry within the structural type. Thus, as has been shown elsewhere [Priestley,2000] a slender bridge column has lower ductility capacity than a squat column with the same section properties and axial load level, provided shear is controlled. With ductile frames designed for a beam-sway inelastic mechanism, the ductility capacity is influenced by the ratio of contribution to the yield displacement resulting from beam flexure, and all other elastic mechanism, including column flexure and shear deformation, and joint shear deformation. Further, for almost all frame buildings, and most cantilever wall buildings, code drift limits will restrict actual ductility levels to values that are lower than those corresponding to the code behaviour factors, and hence the specified behaviour factors are irrelevant to design [Priestley,2003]. This essentially requires an extra level of iteration in a conventional design using initial stiffness, since the strength, and hence stiffness cannot be fully defined until the behaviour factor appropriate for the code drift limit can be determined.

4.3 Displacement-Equivalence Rules for Initial Stiffness Design.

A key tenet of initial-stiffness based seismic design is that displacement of the designed inelastic structure can be related to the elastic displacement of the initial-stiffness model by a displacement-equivalence rule. The equivalence model [Newmark and Hall, 1987] implies that the equal-displacement approx-imation applies for periods greater than about 0.5 sec, with equal acceleration applying at T=0. At intermediate periods the equal acceleration approximation, which implies inelastic displacements that are larger than the elastic displacements is adopted. This equivalence rule, or variants of it, have been widely adopted in initial-stiffness design (see [Chopra and Goel, 2001,e.g.], though some more recent models imply that in the moderate-period range (0.5s $< T <$ 3s), inelastic displacements are less than elastic displacements.

In reviewing displacement-equivalence rules it is important to recall that it is common to specify a level of elastic damping in inelastic time-history analysis (ITHA) to represent damping in the initial stages of response, before hysteretic damping is activated. This is normally specified as a percentage, typically 5%, of critical damping. There are a number of ways this damping could be defined, as discussed subsequently, but the principal difference is whether the damping force is related to the initial or tangent stiffness. In initial-stiffness elastic damping the damping force is always related to the initial stiffness and the instantaneous velocity, while in tangent-stiffness damping, the damping force reduces when the stiffness reduces as a consequence of onset of inelastic response.

Typically research papers reporting results on SDOF ITHA to develop displacement-equivalence rules have assumed initial-stiffness damping. It is of interest to examine the relevance and consequences of this assumption.

There are three main reasons for incorporating initial elastic damping in ITHA:

1. The assumption of linear elastic response at force levels less than yield. Many hysteretic rules, including the common ones used to calibrate displacement-equivalence rules (e.g. elasto-plastic, bilinear, Takeda) make this assumption and therefore do not represent the non-linearity, and hence hysteretic damping within the elastic range , unless additional elastic damping is provided.
2. Foundation damping: soil flexibility, nonlinearity and radiation damping are not normally in structural time-history analyses, and may provide additional damping to the structural response.
3. Non-structural damping: hysteretic response of non-structural elements in a building may result in an effective additional damping force.

Discussing these reasons in turn, it is noted that hysteretic rules are generally calibrated to experimental structural response in the inelastic phase of response. Therefore additional elastic damping should not be used in the post-yield state to represent structural response except when the structure is loading and unloading elastically. Clearly tangent-stiffness damping would thus be more appropriate for modeling structural response.

If the structure deforms with perfect plasticity in the post-yield phase, then foundation forces will remain constant, and foundation damping will cease. Again, tangent-stiffness damping will best model this.

It is conceivable that the non-structural damping force is displacement-dependent rather than force-dependent, and hence a constant damping coefficient may be appropriate for the portion of "elastic" damping that is attributable to non-structural action. However, calculations based on typical non-structural elements in office buildings indicate that an upper bound of about 0.5% equivalent viscous damping related to the structural response seems reasonable [Priestley and Grant, 2006].

The choice between initial-stiffness and tangent-stiffness damping has considerable significance to displacement -equivalence rules. It is shown elsewhere (Priestley and Grant, 2006) that displacements predicted by ITHA using initial stiffness or tangent-stiffness elastic damping representation can differ by as much as 50%, with the difference increasing at low initial periods, high ductility levels, and with hysteresis rules with low inherent energy absorption.

4.4 Distribution of Required Strength in Force-Based Seismic Design

In force-based seismic design, the initial stiffness estimate of different structural members is also used to distribute the total required strength (i.e. the base shear) to the various members. This can result in undesirable and illogical consequences. We consider a few cases in the following examples:

Bridge with varying column heights

Consider a bridge, crossing a valley, which as a consequence of the terrain has piers of different heights, as shown in Figure 3. Under longitudinal response the deflections at the top of the piers will be essentially equal. Assuming a pinned connection between the pier tops and the superstructure (or alternatively, fixed connections and a rigid superstructure), initial-stiffness design will allocate the seismic design force between the piers in proportion to their elastic stiffnesses. If the columns have the same cross-section dimensions, as is likely to be the case for architectural reasons, the design forces in the columns, V_A, V_B, and V_C will be in inverse proportion to H_{A3}, H_{B3}, and H_{C3} respectively, since the stiffnesses are given by

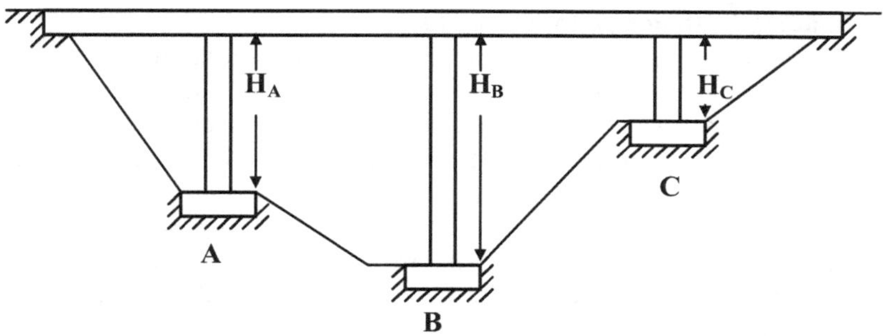

Figure 3 Bridge with Unequal Pier Heights.

$$K_i = C_1 EI_{i,eff} / H_i^3 \quad \textbf{(9)}$$

where C_1 is a constant dependent on the degree of fixity at the pier top, and $I_{i,eff}$ is the effective cracked-section stiffness, typically taken as $0.5I_{gross}$ for all columns. The consequence of this design approach is that the design moment at the bases of the piers will be

$$M_{B,i} = C_2 V_i H_i = C_1 C_2 EI_{i,eff} / H_i^2 \quad \textbf{(10)}$$

where again C2 depends on end fixity conditions. Thus the design moments, based on initial stiffness design, will be in inverse proportion to the square of the column height.

There are a number of consequences of this force distribution. First it is recalled that the yield curvature of a concrete section is independent of the strength, as noted above. Hence the yield displacements of the columns will be proportional to H_{i2}, regardless of the strength allocated to them. If column *B* has a height 2.5 times that of column *C*, as suggested in Figure 3, then the yield displacement of column *B* will be 6.25 times that of Column *C*. Clearly the ductility demands of the piers will be grossly different, and assessment of a characteristic structural ductility factor, and hence of a force-reduction or behaviour factor will be problematic.

Second, the shortest piers will be allocated much higher flexural reinforcement contents than the longer piers. This has several undesirable effects. Allocating more flexural reinforcement to the short piers will increase there elastic stiffness even further, with respect to the longer piers, since the section stiffness is proportional to strength, as noted above. A re-design, should be thus carried out with the improved estimates of column stiffness, which would result in a still higher percentage of total seismic resisting force being allocated to the shorter piers. Allocating a large proportion of the total seismic design force to the short piers increases their vulnerability to shear failure. Further, the displacement capacity of heavily reinforced columns is reduced as the reinforcement ratio increases [Priestley et al, 1996], and hence the initial stiffness approach will tend to reduce displacement capacity of the structure as a whole.

Structural wall buildings with unequal wall lengths

A similar problem with initial stiffness structural characterization to that discussed for bridges above occurs when buildings are provided with cantilever walls of different length providing seismic resistance in a given direction, or when a mixed seismic-resisting system comprised of walls and frames is adopted. We consider just the structural wall building here, with reference to Figure 4.

An initial-stiffness design will again allocate the base shear force between the walls in proportion to their elastic stiffness, which in this case is in proportion to $l_{w,i3}$. The assumption is that by doing so, the walls can be forced to yield at the same displacement. This is, in fact, incorrect, since the yield curvature is constant for a given section, as noted above, and is given by Equation (7), with $h = l_{w,i}$. Thus the yield displacement of the shorter walls will always be greater than that of the longer walls, by the same proportion, regardless of the allocated strength, as shown in Figure 4. The ductility demands on the walls will thus differ, and the same problem as faced in the previous example, that of deciding an appropriate force-reduction factor must be faced here.

As noted, initial-stiffness design will allocate the design base shear force to the walls in proportion the cube of the wall length. Strength is unnecessarily, and unwisely concentrated in the stiffest elements, underutilizing the more flexible members. Reinforcement ratios for the walls will be roughly proportional to the wall lengths. Again, a reassessment of the force distribution between walls based on designed reinforcement content and initial stiffness would result in a relative increase in the stiffness of the longer walls and hence an allocation of shear force larger than in proportion to length cubed.

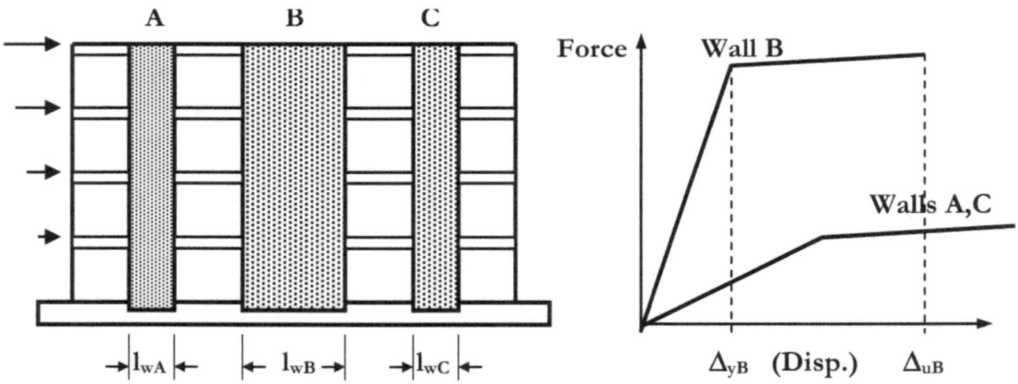

Figure 4 Building with Unequal Length Cantilever Walls.

Bridge with dual load paths

A final example is related to structures which possess more than one seismic load path, one of which remains elastic while the others respond inelastically at the design earthquake level. A common example is the bridge of Figure 5(a), subjected to transverse seismic excitation, as suggested by the double-headed arrows. Primary seismic resistance is provided by bending of the piers, which are designed for inelastic response. However, if the abutments are restrained from lateral displacement under transverse response, superstructure bending also develops under transverse excitation. Current seismic design philosophy requires that the superstructure responds elastically. The consequence is that a portion of the seismic inertia forces developed in the superstructure are transmitted to the abutments by column bending (path 1) and the remainder is transmitted to the abutments by superstructure flexure (path 2). With initial-stiffness characterization, an elastic analysis is carried out and the relative elastic stiffnesses of the two paths are as indicated by the two broken lines of Figure 5(b), which indicates that column flexure carries most of the seismic force. A force-reduction factor is then applied, and the design forces corresponding to yield displacement are determined.

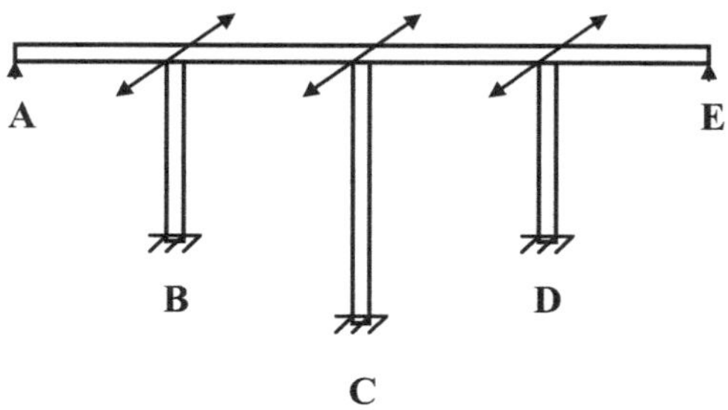

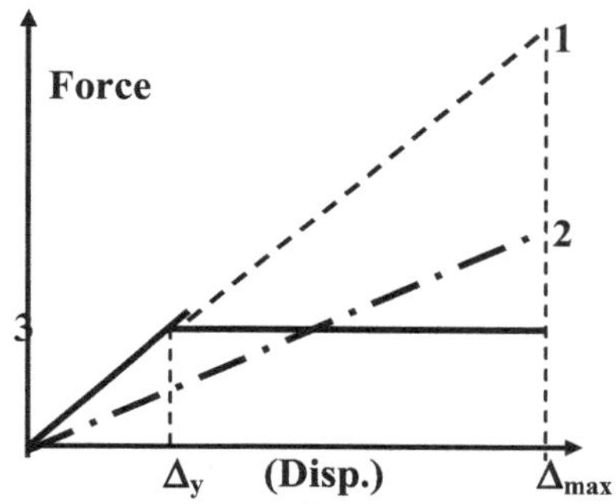

Figure 5 Bridge with Dual Load Paths Under Transverse Seismic Excitation.

The inelastic response of the combined resistance of the columns is now shown by the solid line (path 3), and on the basis of the equal displacement approximation it is imagined that the maximum displacement is Δ_{max}, the value predicted by the elastic analysis. If the lateral strength of the superstructure is designed on the basis of the force developed in path 2 at the column yield displacement, it will be seriously under-designed, since the forces in this path, which are required to be within the elastic range, continue to rise with increasing displacement. Thus the bending moment in the superstructure, and the abutment reactions are not reduced by column hinging, and a force-reduction factor should not be used in their design.

It is also probable that the maximum response displacement will differ significantly from the initial elastic estimate, since at maximum displacement, the damping of the system will be less than expected, as hysteretic damping is only associated with load path 3, which carries less than 50% of the seismic force at peak displacement response. This may cause an increase in displacements. On the other hand, the higher strength associated with the increased post-yield stiffness of load path 2 may result in reduced displacement demand. Initial-stiffness based analysis and the force-reduction factor approach give no guidance to these considerations.

5 Conclusions

In this chapter we have shown that there are a number of problems inherent in force-based seismic design. These include the determination of the required strength based on estimated stiffnesses which in fact depend on the final allocated strength, the assumption that behaviour factors can be defined for a structural class and material, and the distribution of strength between members based on relative initial stiffness of the members. We will see in the next chapter that these problems disappear if we use a displacement-based seismic design approach based on secant stiffness to maximum design displacement.

6 References

Chopra, A.K. and Goel, R.K. (2001). Direct Displacement-Based Design: Use on Inelastic vs. Elastic Design Spectra. *Earthquake Spectra 17(1), pp 47-65*

Newmark, N. and Rosenbleuth, E. (1966). Seismic Design. *Prentice Hall*

Newmark, N.M. and Hall, W.J. (1987). Earthquake Spectra and Design. *EERI Monograph,103 pp.*

Park, R. and Paulay, T. (1976). Reinforced Concrete Structures. *John Wiley and Sons, New York*

Paulay, T. and Priestley, M.J.N. (1992). Seismic Design of Reinforced Concrete and Masonry Structures *John Wiley and Sons, New York*

Priestley, M.J.N., Seible, F. and Calvi, G.M. (1996). Seismic Design and Retrofit of Bridges. *John Wiley and Sons, New York*

Priestley, M.J.N. (1998). Brief Comments on Elastic Flexibility of Reinforced Concrete Frames, and Significance to Seismic Design. *Bulletin, NZ National Society for Earthquake Engineering, 31(3)*

Priestley, M.J.N. (2000). Peformance Based Seismic Design. *Keynote address, Proceedings of the 12th World Conference on Earthquake Engineering*

Priestley, M.J.N. (2003). Myths and Fallacies in Earthquake Engineering, Revisited. The Mallet Milne Lecture. *IUSS Press, Pavia, Italy*

Priestley, M.J.N. and Grant, D.N. (2005). Viscous Damping in Seismic Design and Analysis. *J. Earthquake Engineering, 9(Sp. Issue 2), pp. 229-255*

Priestley, M.J.N., Calvi, G.M. and Kowalsky, M.J. (2007). Direct Displacement-Based Seismic Design of Structures. *IUSS Press, Pavia (in press)*

Fundamentals of Direct Displacement-Based Seismic Design and Assessment *

Nigel Priestley[1]

[1] Emeritus Professor of Structural Engineering, University of California, San Diego

Abstract: This chapter presents the fundamentals of the new seismic design method known as direct displacement based design. It is a simple design approach where the multi-degree-of-freedom structure is characterized the secant stiffness and equivalent elastic damping of an equivalent single-degree-of-freedom structure. Design is based on achieving a specified displacement limit state – defined either by material strain limits or non-structural drift limits – under the design level seismic intensity, rather than using these limits merely as upper bounds of acceptable behaviour. The characterization of the structure by secant stiffness avoids the many problems inherent in force-based design where initial stiffness is used to determine an elastic period, and forces are distributed between members in proportion to elastic stiffness.

1 Introduction

An alternative seismic design philosophy, now known as Direct Displacement-Based Design (DDBD) was introduced in (Priestley,1993). Although this is but one of a number of seismic design procedures recently developed and jointly termed "Performance Based Seismic Design", it differs in significant details from the other methods. It is our contention that DDBD is more intellectually satisfying than the alternatives, and is best equipped to address the deficiencies of conventional force-based design. It has also now been developed in rather more complete form than the other methods, and has been applied to a wider category of structures. Finally, and perhaps of greatest importance, it is simpler to apply, and better suited to incorporation in design codes. For these reasons, the fundamental considerations of the method are outlined in this chapter, in considerably more detail than was possible in the original development.

The fundamental difference from force-based design is that DDBD characterizes the structure to be designed by a single-degree-of-freedom (SDOF) representation of performance at peak displacement response, rather than by its initial elastic characteristics. This is based on the Substitute Structure approach pioneered by Sozen and co-workers (Gulkan and Sozen,1974; Shibata and Sozen, 1976).

The design approach attempts to design a structure which would achieve, rather than be bounded by, a given performance limit state under a given seismic intensity, essentially resulting in uniform-risk structures, which is philosophically compatible with the uniform-risk seismic spectra incorporated in most design codes. The design procedure determines the strength required at designated plastic hinge locations to achieve the design objectives in terms of defined displacement objectives. It must then be combined with capacity design procedures to ensure that plastic hinges occur only where intended, and that non-ductile modes of inelastic deformation do not develop.

This chapter deals with fundamental aspects of the approach that are common to all materials and structural systems. More complete information, providing detailed application to different structural systems is in the final stages of preparation (Priestley et al,2007).

2 Basic Formulation of the method

The design method is illustrated with reference to Figure 1, which considers a SDOF representation of a frame building (Figure 1(a)), though the basic fundamentals apply to all structural types. The bilinear envelope of the lateral force-displacement response of the SDOF representation is shown in Figure 1(b). An initial elastic stiffness K_i is followed by a post yield stiffness of r_{Ki}.

While force-based seismic design characterizes a structure in terms of elastic, pre-yield, properties (initial stiffness K_i, elastic damping), DDBD characterizes the structure by secant stiffness K_e at maximum displacement $_d$ (Figure 1(b)), and a level of equivalent viscous damping ξ, representative of the combined elastic damping and the hysteretic energy absorbed during elastic response. Thus, as shown in Figure 1(c), for a given level of ductility demand, a structural steel frame building with compact members will be assigned a higher level of equivalent viscous damping than a reinforced concrete frame building designed for the same level of ductility demand, as a consequence of "fatter" hysteresis loops.

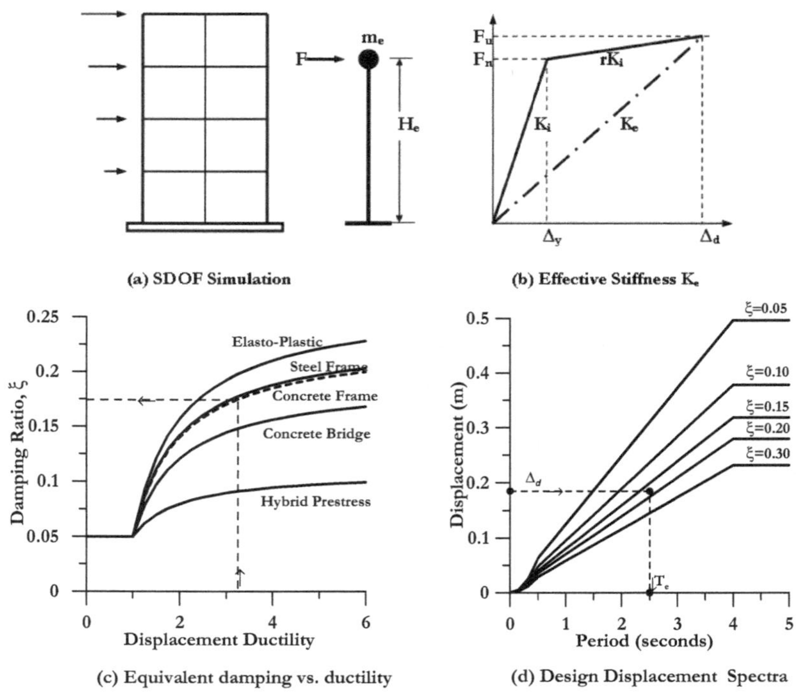

Figure 1 Fundamentals of Direct Displacement-Based Design.

With the design displacement at maximum response determined, as discussed above, and the corresponding damping estimated from the expected ductility demand (Figure 1(c)), the effective period T_e at maximum displacement response can be read from a set of displacement spectra for different levels of damping, as shown in the example of Figure 1(d). The effective stiffness K_e of the equivalent SDOF system at maximum displacement can be found by inverting the normal equation for the period of a SDOF oscillator, to provide Equation (1):

$$K_e = 4\pi^2 m_e / T_e^2 \qquad \textbf{(1)}$$

where m_e is the effective mass of the structure participating in the fundamental mode of vibration. From Figure 1(b), the design lateral force, which is also the design base shear force is thus:

$$F = V_B = K_e \Delta_d \qquad \textbf{(2)}$$

The design concept is thus very simple. Such complexity that exists relates to determination of the "substitute structure" characteristics, the determination of the design displacement, and development of design displacement spectra. Careful consideration is however necessary for the distribution of the design base shear force V_B through the structure, and the analysis of the structure under the distributed seismic force.

3 Single-Degree-of-Freedom (SDOF) Structures

3.1 Design Displacement for a SDOF structure

The design displacement will depend on the limit state being considered, and whether structural or non-structural considerations are more critical. For any given limit state (e.g. serviceability limit state, damage control limit state, etc.) structural performance will be governed by limiting material strains, since damage is strain-related for structural elements. Damage to non-structural element can be generally considered drift-related.

It is comparatively straightforward to compute the design displacement from strain limits. Consider the vertical cantilever structure of Figure 2(a). The most realistic structure conforming to the assumptions of a SDOF approximation is a regular bridge under transverse excitation. Two possible reinforced concrete sections, one circular and one rectangular are shown in Figure 2(b). The strain profile at maximum displacement response is shown together with the sections. Maximum concrete compression strains ε_c and reinforcement tensile strain εs are developed. The limit-state strains are ε_{cm} and ε_{sm} for concrete compression and steel tension respectively. These will not generally occur simultaneously in the same section, since the neutral axis depth c is fixed by the reinforcement ratio, and the axial load on the section. Consequently there are two possible limit state curvatures, based on the concrete compression and the reinforcement tension respectively:

$$\phi_{mc} = \varepsilon_{cm} / c \qquad \textbf{(3a)}$$

$$\phi_{ms} = \varepsilon_{sm}/(d-c) \quad \textbf{(3b)}$$

The lesser of ϕ_{mc} and ϕ_{ms} will govern the structural design. The design displacement can now be estimated as:

$$\Delta_{ds} = \Delta_y + \Delta_p = \phi_y H^2/3 + (\phi_m - \phi_y) L_p H \quad \textbf{(4)}$$

where ϕ_m is the lesser of ϕ_{mc} and ϕ_{ms}, ϕ_y and Δ_y are the yield curvatures and displacement, respectively, H is the column height (see Figure 2) and L_p is the plastic hinge length.

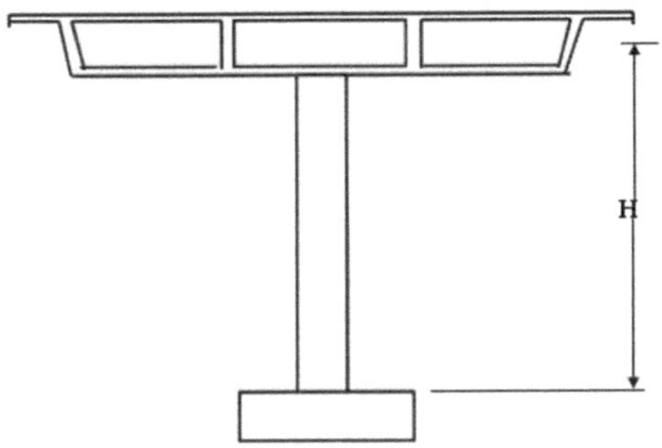

(a) Cantilever Bridge Column

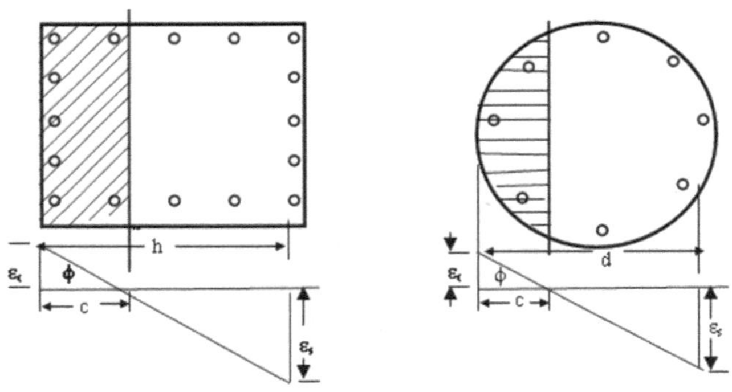

(b) Column Sections and Limit State Strains

Figure 2. Limit State Strain Profiles for a Cantilever Bridge Column.

If the limit state has a non-structural drift limit θ_c the displacement given by Equation (4) must be checked against

$$\Delta_{d\theta} = \theta_c H \qquad \textbf{(5)}$$

The lesser of the displacements given by Equations (4) and (5) is the design displacement.

Note that an alternative design approach would be to design the structure for a specified drift, and then determine the details to ensure the strain limits are achieved. For example, the limit concrete strain for the damage-control limit state can be determined (Priestley et al., 1996) from the transverse reinforcement details by:

$$\varepsilon_{cm} = 0.004 + 1.4\rho_s f_{yh} \varepsilon_{su} / f_{cc}^{'} \qquad \textbf{(6)}$$

where ρ_s, f_{yh}, and ε_{su} are the volumetric ratio, yield strength and strain at ultimate stress of the transverse reinforcement, and f'_{cc} is the compression strength of the concrete confined by the transverse reinforcement.

Thus the concrete strain corresponding to the drift limit can be determined by inverting Equations (4) and (3a), and the required amount of transverse reinforcement calculated by inverting Equation (6). This simplifies the design process.

3.2 Yield Displacement

For a SDOF vertical cantilever, the yield displacement is required for two reasons. First, if structural considerations define the limit displacement (Equation (3)), the yield displacement and yield curvature must be known. Second, in order to calculate the equivalent viscous damping, the displacement ductility μ_Δ, which depends on the yield displacement, must be known.

It has been shown (Priestley, 2003) that for reinforced concrete (and masonry) members, the yield curvature is essentially independent of reinforcement content and axial load level, and is a function of yield strain and section depth alone. A summary of expressions for the yield curvature and drift is given in Equation (7).

Circular concrete column:	$\phi_y = 2.25\varepsilon_y / D$	**(7a)**
Rectangular concrete column:	$\phi_y = 2.10\varepsilon_y / h_c$	**(7b)**
Rectangular concrete wall:	$\phi_y = 2.00\varepsilon_y / l_w$	**(7c)**
Symmetrical steel section:	$\phi_y = 2.10\varepsilon_y / h_s$	**(7d)**
Flanged concrete beam:	$\phi_y = 1.70\varepsilon_y / h_b$	**(7e)**

where ε_y is the yield strain of the flexural reinforcement ($= f_y/E_s$), and D, h_c, l_w, h_s and h_b are the section depths of the circular column, rectangular column, rectangular wall, steel section and flanged concrete beam sections respectively. Note that Equation (7) gives the curvature at the yield of the equivalent bilinear approximation to the moment-curvature curve. As such it is a useful reference value when using bilinear force-displacement modelling.

For a SDOF vertical cantilever, such as a bridge pier, or a low rise cantilever wall, the yield displacement can be satisfactorily approximated for design purposes by:

$$\Delta_y = \phi_y H^2 / 3 \quad \textbf{(8)}$$

For reinforced concrete and structural steel frames the yield drift can be developed from the yield curvature expressions of Equations (7) as:

Reinforced concrete frame: $$\theta_y = 0.5\varepsilon_y l_b / h_b \quad \textbf{(9a)}$$

Structural steel frame: $$\theta_y = 0.65\varepsilon_y l_b / h_b \quad \textbf{(9b)}$$

where l_b is the beam span, and h_b is the concrete or steel beam depth. It will be noted that the yield drifts, and hence the yield displacements of reinforced concrete and structural steel frames with similar geometries differ only by 30%, and that concrete frames are typically stiffer.

3.3 Equivalent Viscous Damping

The design procedure requires relationships between displacement ductility and equivalent viscous damping, as shown in Figure 1(c). The damping is the sum of elastic and hysteretic damping. Normally the elastic damping is taken as 5%, and hence the design damping is

$$\xi_d = 0.05 + \xi_{hyst} \quad \textbf{(10)}$$

where the hysteretic damping ξ_{hyst} is related (but not equal) to the area within the stabilized hysteretic force-displacement response (see Figure 3) as:

$$\xi_{hyst} = K.\frac{A_h}{2\pi F_m \Delta_m} \quad \textbf{(11)}$$

In Equation (11), Ah is the area with one complete cycle of stabilized force-displacement response, F_m and Δ_m are the maximum force and displacement achieved in the stabilized loops, and K is a constant calibrated from inelastic time-history analyses. Note that the damping given by Equation (10) and (11) is expressed as a ratio of the critical damping coefficient, and is related to the secant stiffness K_e to maximum response. It is thus compatible with the assumptions of structural characterization by stiffness and damping at peak response.

For general design purposes the following equations may be used to estimate the total damping:

Concrete Wall Building, Bridges (TT): $\xi_{eq} = 0.05 + 0.444\left(\frac{\mu - 1}{\mu\pi}\right)$ **(12a)**

Concrete Frame Building (TF): $\xi_{eq} = 0.05 + 0.565\left(\frac{\mu - 1}{\mu\pi}\right)$ **(12b)**

Steel Frame Building (RO): $\xi_{eq} = 0.05 + 0.577\left(\frac{\mu - 1}{\mu\pi}\right)$ **(12c)**

Hybrid Prestressed Frame (FS, β=0.35): $\xi_{eq} = 0.05 + 0.186\left(\frac{\mu - 1}{\mu\pi}\right)$ **(12d)**

Friction Slider (EPP): $\xi_{eq} = 0.05 + 0.670\left(\frac{\mu - 1}{\mu\pi}\right)$ **(12e)**

Bilinear Isolation System (BI, r=0.2): $\xi_{eq} = 0.05 + 0.519\left(\frac{\mu - 1}{\mu\pi}\right)$ **(12f)**

In Equations (12) $\mu = \Delta_d/\Delta_y$ is the displacement ductility at the design displacement.

3.4 Design Displacement Spectra

Figure 1(d) shows a typical design displacement spectra set for damping levels between 5% and 30% of critical damping. This was generated from the design 5% acceleration spectrum for firm ground and a peak ground acceleration of 0.4g, according to Eurocode EC8. It will be noted that for periods between 0.5 and 4.0 seconds the 5% displacement spectrum is linear. At 4.0 seconds, a displacement cut off is imposed, and at higher periods, the design displacement is considered to be equal to the 4.0 second value. This is in recognition of the fact that peak response displacements from recorded accelerograms tend to occur between 1.5 and 4.0 seconds, and then decrease at higher periods. Note that EC8 imposes the displacement cut-off at 2.0 seconds, but this is believed to be non-conservative. The linear relationship between period and displacement between 0.5 and 4.0 seconds is a result of the inverse relationship between response acceleration and period in the EC8 (and in most other codes) acceleration spectrum.

Ideally, design displacement spectra should be generated separately from the acceleration spectrum. However, adequate results can be obtained assuming sinusoidal relationships at peak acceleration and displacement response:

$$\Delta_{T,5} = \frac{T^2}{4\pi^2} = a_{T,5}g \qquad \textbf{(13)}$$

where $\Delta_{T,5}$ and $\alpha_{T,5}$ are the response displacement and accelerations for 5% damping, at period T, and g is the acceleration of gravity. For normal accelerograms measured at least 10 km from the fault rupture, the displacements corresponding to a different level of damping, ξ, can be related to the displacement for 5% damping by the EC8 expression.

$$\Delta_{T,\xi} = \Delta_{T,5} \cdot \left(\frac{7}{2+\xi} \right)^{0.5} \quad \textbf{(14)}$$

Equation (14) uses the EC8 expression from the 1998 version of the code. The latest version of the expression, in the current code is overly conservative and does not agree with analytical results. In the near-field region, the influence of velocity pulses may reduce the effectiveness of damping (and hysteretic energy absorption). In these regions, Equation (14) is likely to be non-conservative, and it is suggested that Equation (15) be used:

$$\Delta_{T,\xi} = \Delta_{T,5} \cdot \left(\frac{7}{2+\xi} \right)^{0.25} \quad \textbf{(15)}$$

Equations (14) and (15) are compared for a range of different levels of damping in Figure 3, in dimensionless form, related to the displacement for 5% damping at a period of $T = 4$ sec, using the EC8 acceleration spectral shape.

For the velocity pulse type ground motion (Figure 3(b)), the influence of damping in reducing the displacement is much less pronounced than for the "normal" ground motion case of Figure 3 (a). The consequence will be that for a given design displacement and damping, the effective period at peak response for the velocity-pulse ground motion will be smaller than for normal ground motion (follow the dashed lines in Figure 3). From Equation (1) the required effective stiffness will thus be increased, and from Equation (2) the required strength increases proportionately, for the velocity pulse case compared to the normal case. Thus the effects of the special character of velocity pulse ground motions can be directly considered in direct displacement-based design, whereas "correction factors" have to applied in force-based design.

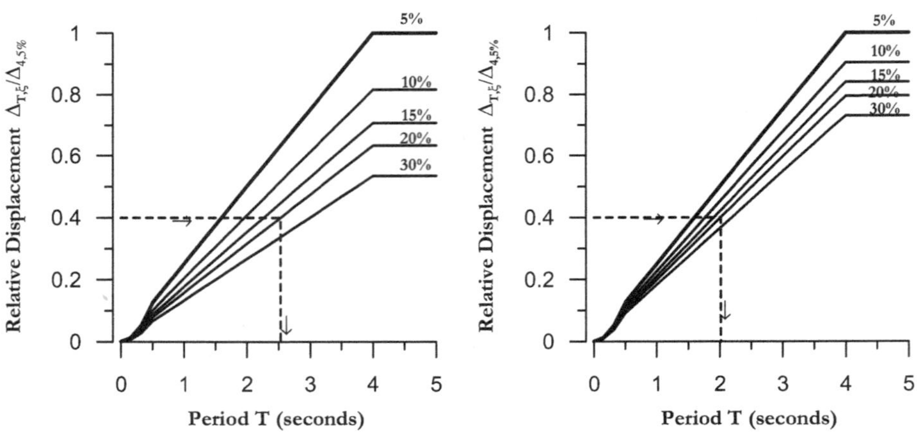

(a) "Normal" (Equation 14) (b) Velocity Pulse (Equation 15)

Figure 3. Influence of Damping on Displacement Response

3.5 Design Base Shear Equation

It will be clear that the approach described above can be simplified to a single design equation, once the design displacement and damping have been determined. As noted above, the displacement spectra are typically linear with effective period. The small non-linearity at low periods is unlikely to be significant for displacement-based designs, since it is the effective period at peak displacement response, approximately $\mu^{0.5}$ times the elastic period, that is of relevance. In Figures 1(d) and 3 the displacements are capped at a period of *4* seconds. Let $\Delta_{P,5}$ be the displacement at the cap period T_P (e.g. $T_P = 4$ seconds in Figure 1d) for the displacement spectrum corresponding to 5% damping. For a design displacement of Δ_d and design damping ξ, the effective period is, from Figure 3 and Eqs. (14) and (15):

$$T_e = T_P \cdot \frac{\Delta_d}{\Delta_{P,5}} \cdot \left(\frac{2+\xi}{7}\right)^{\beta} \quad \textbf{(16)}$$

where $\beta = 0.5$ and 0.25 for normal and velocity pulse conditions respectively. From Equation (1) the effective stiffness at peak response is thus:

$$K_e = \frac{4\pi^2 m_e}{T_P^2} \cdot \frac{\Delta_{P,5}^2}{\Delta_d^2} \cdot \left(\frac{7}{2+\xi}\right)^{2\beta} \quad \textbf{(17)}$$

Finally from Equations (2) and (17):

$$V_B = K_e \Delta_d = \frac{4\pi^2 m_e}{T_P^2} \cdot \frac{\Delta_{P,5}^2}{\Delta_d} \cdot \left(\frac{7}{2+\xi}\right)^{2\beta} \quad \textbf{(18)}$$

4 Multi-Degree of Freedom Structures

For multi-degree-of-freedom (MDOF) structures the initial part of the design process requires the determination of the characteristics of the equivalent SDOF substitute structure. The required characteristics are the equivalent mass, the design displacement, and the effective damping. When these have been determined, the design base shear for the substitute structure can be determined. The base shear is then distributed between the mass elements of the real structure as inertia forces, and the structure analyzed under these forces to determine the design moments at locations of potential plastic hinges.

4.1 Design Displacement

The characteristic design displacement of the substitute structure depends on the limit state displacement or drift of the most critical member of the real structure, and an assumed displacement shape for the structure. This displacement shape is that which corresponds to the inelastic first-mode at the design level of seismic excitation. Thus the changes to the elastic first-mode shape resulting from local changes to member stiffness caused by inelastic action in plastic hinges are taken into account at the beginning of the design. Representing the displacement by the inelastic rather than the elastic first-mode shape is consistent with characterizing the structure by its secant stiffness to maximum response. In fact, the inelastic and elastic first-mode shapes are often very similar.

The design displacement (generalized displacement coordinate) is thus given by:

$$\Delta_d = \sum_{i=1}^{n}\left(m_i\Delta_i^2\right) / \sum_{i=1}^{n}\left(m_i\Delta_i\right) \quad \textbf{(19)}$$

where m_i and Δ_i are the masses and displacements of the n significant mass locations respectively. For multi-storey buildings, these will normally be at the n floors of the building. For bridges, the mass locations will normally be at the top of the columns, but the superstructure mass may be discretized to more than one mass per span to improve validity of simulation.

Where strain limits govern, the design displacement, or drift, of the critical member can be determined using a similar approach to that outlined in Section 2. Similar conclusions apply when code drift limits apply. For example, the design displacement for frame buildings will normally be governed by structural or non-structural drift in the beams in the lower levels of the building. For a bridge, it will normally be governed by the plastic rotation of the shortest column. With a knowledge of the displacement of the critical member and the design displacement shape (discussed further in the following section), the displacements of the individual masses are given by:

$$\Delta_i = \phi_i \cdot \left(\frac{\Delta_c}{\phi_c}\right) \quad \textbf{(20)}$$

where ϕ is the inelastic mode shape, and Δ_c is the design displacement at the critical mass.

4.2 Displacement Shapes

Frame buildings

For regular frame buildings, the following equations, though approximate, have been shown to be adequate for design purposes:

for n<4: $$\phi_i = H_i / H_n \quad \textbf{(21a)}$$

$4 < n < 20$: $$\phi_i = \frac{H_i}{H_n} \cdot \left[\frac{16 - 0.5 * \frac{H_i}{H_n} \cdot (n-4)}{16 - 0.5(n-4)}\right] \quad \textbf{(21b)}$$

n > 20:
$$\phi_i = \frac{2H_i}{H_n} \cdot \left(1 - 0.5\frac{H_i}{H_n}\right) \qquad \textbf{(21c)}$$

In Equations (21) H_i and H_n are the heights of level i, and the roof (level n) respectively. Shapes for different numbers of stories, resulting from Equations (21) are shown in Figure 4.

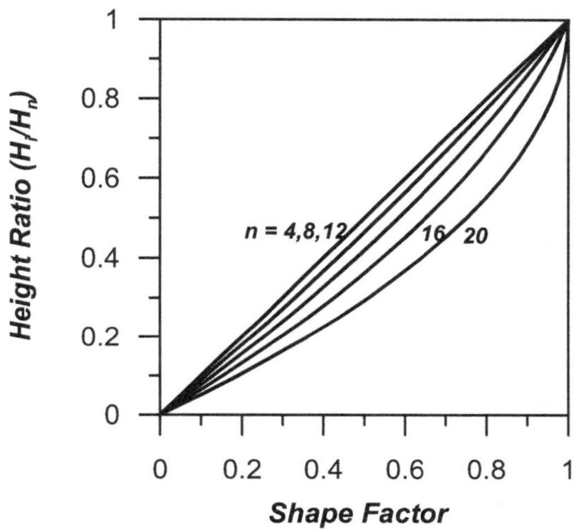

Figure 4. Design Displacement Profiles for Frames

Cantilever wall buildings:
For cantilever wall buildings the maximum drift will occur in the top storey. The value of this drift may be limited by the code maximum drift limit, or by the plastic rotation capacity of the base plastic hinge. From Equation (19), assuming a triangular distribution of curvature with height, as shown in Figure 5, the yield drift at the top of the wall will be:

$$\theta_{yn} = \phi_y H_n / 2 = 1.0\varepsilon_y H_n / l_w \qquad \textbf{(22)}$$

As a reasonable approximation, the plastic rotation may be concentrated at the wall base. Hence the critical drift at the top of the wall will be:

$$\theta_{dn} = \theta_{yn} + \theta_{pn} = 1.0\varepsilon_y H_n / l_w + (\phi_m - 2.0\varepsilon_y / l_w)L_p \leq \theta_c \qquad \textbf{(23)}$$

where θ_{pn} is the plastic rotation at the top of the wall corresponding to the design limit.

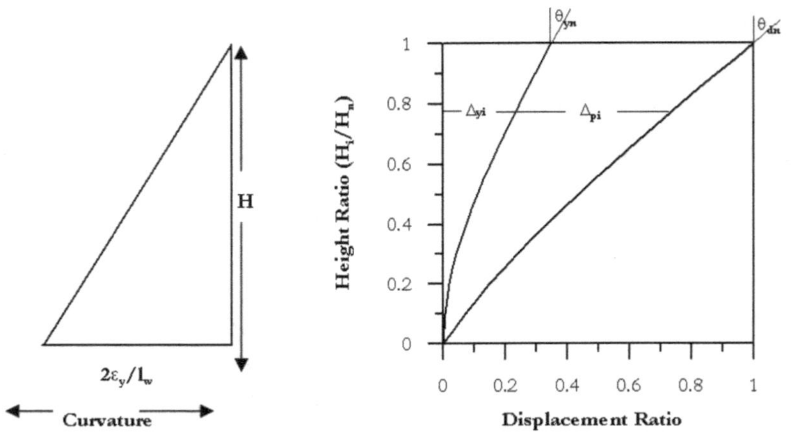

Figure 5. Yield and Design Displacement Profiles for Walls

state, ϕ_m is the corresponding base curvature, and L_p is the plastic hinge length at the wall base, given by the larger of:

$$L_p = 0.054H_n + 0.022f_y \cdot d_{bl} \quad \textbf{(24a)}$$

$$L_p = 0.2l_w + 0.03H_n \quad \textbf{(24b)}$$

The yield displacement at height H_i is given by:

$$\Delta_{yi} = \frac{\varepsilon_y}{l_w} H_i^2 \left(1 - \frac{H_i}{3H_n}\right) \quad \textbf{(25a)}$$

If the roof drift from Equation (22) is less than the code drift limit θ_c, then the design displacement profile is given by:

$$\Delta_i = \Delta_{yi} + \Delta_{pi} = \frac{\varepsilon_y}{l_w} H_i^2 \left(1 - \frac{H_i}{3H_n}\right) + \left(\phi_m - \frac{2\varepsilon_y}{l_w}\right) L_p H_i \quad \textbf{(25b)}$$

If the code drift limit governs the roof drift, the design drift is given by:

$$\Delta_i = \Delta_{yi} + (\theta_c - \theta_{yn})H_i = \frac{\varepsilon_y}{l_w} H_i^2 \left(1 - \frac{H_i}{3H_n}\right) + \left(\theta_c - \frac{\varepsilon_y H_n}{l_w}\right) H_i \quad \textbf{(26)}$$

Although Equation (25) can be manipulated to provide a generalized displacement shape ϕ_i to be compatible with Equation (20), there is little value in so doing, since the full displacement profile must first be found.

Multi-span bridges:
With bridges it is less easy to initially determine a design displacement profile, particularly for transverse seismic response. Figure 6 illustrates two possible bridge configurations out of a limitless potential range. The example of Figure 6(a) has piers of uniform height, while those in Figure 6(b) vary in height. The transverse displacement profiles will depend strongly on the relative column stiffnesses, and more significantly, on the degree of lateral restraint provided at the abutment. For each bridge type, three possible transverse displacement shapes are shown, corresponding to an abutment fully restrained against transverse displacement, a completely unrestrained abutment, and one where the abutment is restrained, but has significant transverse flexibility.

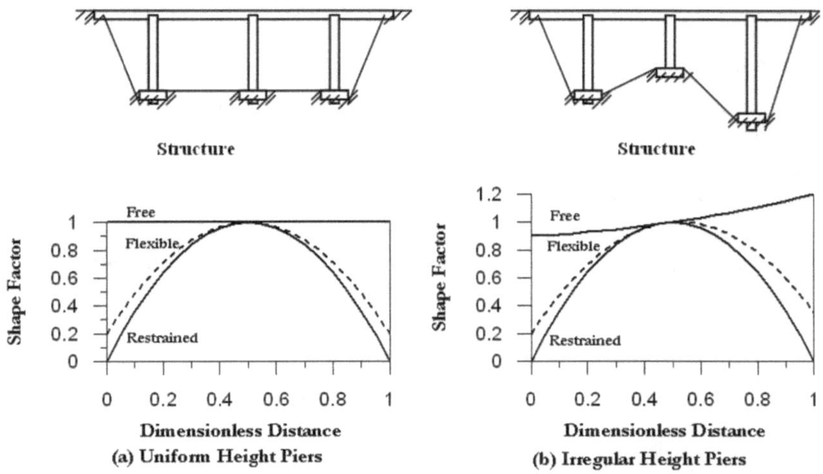

Figure 6. Design Transverse Displacement Profiles for Bridges

For the case of Figure 6(a), the critical pier will be the central one, and with the appropriate displacement profile chosen, Equation (20) can be applied directly. For the irregular bridge of Figure 6(b) the critical pier may not be immediately apparent, and some iteration may be required. Iteration may also be required for the case of finite flexibility of the abutments for the regular bridge example to determine the relative displacements of abutment and the critical pier. Generally a parabolic displacement shape between abutments and piers can be assumed for initial design.

4.3 Effective Mass

From consideration of the mass participating in the first inelastic mode of vibration, the effective system mass for the substitute structure is:

$$m_e = \sum_{i=1}^{n} (m_i \Delta_i) / \Delta_d \tag{27}$$

Typically, the effective mass will range from about 70% of the total mass for multistory cantilever walls to more than 85% for frame buildings of more than 20 storeys. The remainder of the mass participates in the higher modes. Although modal combination rules such as the square-root-sum-of-the-squares (SRSS), or complete modal combination (CQC), may indicate a significant increase in the elastic base shear force over that from the first inelastic mode, there is much less influence on the design base overturning moment. The effects of higher modes are inadequately represented by elastic analysis and direct modal combination, and are better accommodated in the capacity design phase, rather than the preliminary phase of design.

4.4 Effective Damping

The effective damping depends on the structural system and displacement ductility factor, as illustrated in Figure 1(c) and Equations (12). This requires determination of the displacement ductility demand of the substitute structure. This poses few problems, since the design displacement Δ_d has already been determined, from Equation (19). The effective yield displacement Δ_y needs to be interpolated from the profile of displacements at yield (e.g. Equation (25) for cantilever walls, or Equation (21) for frames). This requires a knowledge of the effective height of the substitute structure, which may be taken as:

$$H_e = \sum_{i=1}^{n} (m_i \Delta_i H_i) / \sum_{i=1}^{n} (m_i \Delta_i) \tag{28}$$

The design ductility factor, for use in Equation (12) is then:

$$\mu = \Delta_d / \Delta_y \tag{29}$$

in the usual fashion.

Note that provided reasonable ductility is implied by the design displacement Δ_d, Figure 1(c) and Equations (12) indicate that the damping is not strongly dependent on the ductility, and average values may be adopted. Note also that it has been established (Priestley, 2003) that concrete and masonry structures are much more flexible than normally assumed by designers, and hence code drift limits, rather than displacement ductility capacity tends to govern design. As a consequence, the design ductility, and the effective damping are known at the start of the design process, and no iteration is needed in determining the design base shear force.

When the lateral resistance of a building in a given direction is provided by a number of walls of different length, the ductility demand of each wall will differ, since the yield displacements of the walls will be inversely proportional to the wall lengths (see Equation (25)), while the maximum displacements at design-level response will be essentially equal, subject only to small variations resulting from torsional response. Hence the system damping will need to consider the different effective damping in each wall. A weighted average is appropriate, given by:

$$\xi_e = \sum_{j=1}^{m}\left(V_j \xi_j\right) / \sum_{j=1}^{m} V_j \qquad \textbf{(30)}$$

where V_j and ξ_j are the base shear force and damping of the m walls in a given direction. A rational decision will be to apportion the total base shear force requirement between the walls in proportion to the square of the length. This will result in essentially constant reinforcement ratios between the walls. With wall strength proportional to length squared, Equation (30) may be rewritten as:

$$\xi_e = \sum_{j=1}^{m}\left(l_{wj}^2 \xi_j\right) / \sum_{j=1}^{m} l_{wj}^2 \qquad \textbf{(31)}$$

Conventional force-based design would apportion the base shear between walls in proportion to the cube of wall length, based on the invalid assumption that the walls could be made to achieve simultaneous yield, regardless of length. Since the yield curvature is inversely proportional to length, this is clearly impossible. The consequence of this approach is that the longer walls end up more heavily reinforced for flexure than the shorter walls, which is irrational, and results in further imbalance in elastic stiffness from the design assumption.

Influence of foundation flexibility on effective damping:

Although the influence of foundation flexibility on seismic design can be incorporated into force-based design, albeit with some difficulty, it is rarely considered. Foundation flexibility will increase the initial elastic period, and reduce the ductility capacity corresponding to the strain or drift limit states. It is comparatively straightforward, however, to incorporate the influence of elastic foundation compliance into Direct Displacement-Based Design. If the limit state being considered is strain-limited, then the design displacement will be increased by the elastic displacement corresponding to foundation compliance (this requires a knowledge of the design base moment and shear force, and hence some iteration may be required). If, however, the limit state is defined by code drift limits, there will be no change in the design displacement, thus implying reduced permissible structural deformation.

The second influence relates to the effective damping. Both foundation and structure will contribute to the damping. Consider the force-displacement hysteresis loops of Figure 7, where foundation (Δ_f) and structure (Δ_s) components of the peak response displacement $\Delta_d = \Delta_s + \Delta_f$ have been separated for a cantilever wall building. The equivalent viscous damping for the foundation and for the structure can be separately expressed as:

Foundation:
$$\xi_f = \frac{A_f}{2\pi V_B \Delta_f} \quad \textbf{(32a)}$$

Structure:
$$\xi_s = \frac{A_s}{2\pi V_B \Delta_s} \quad \textbf{(32b)}$$

System:
$$\xi_e = \frac{A_f + A_s}{2\pi V_\mathrm{B}(\Delta_f + \Delta_s)} = \frac{\xi_f \Delta_f + \xi_s \Delta_s}{\Delta_f + \Delta_s} \quad \textbf{(32c)}$$

where A_f and A_s are hysteretic areas within the loops (i.e. energy absorbed per cycle) for foundation and structure respectively. As shown in Figure 7, the hysteretic area of the combined structure/foundation system will be the sum of the two components, and hence the system equivalent damping will be:

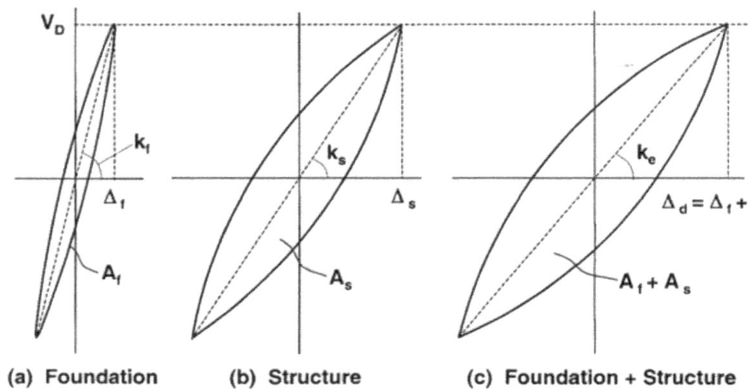

Figure 7. Damping Contributions of Foundation and Structure.

4.5 Distribution of Design Base Shear Force

The principles outlined in the previous sections enable the design base shear to be established for a MDOF system. This base shear force must be distributed as design forces to the various discretized masses of the structure, in order that the design moments for potential plastic hinges can be established. Assuming essentially sinusoidal response at peak response, the base shear should be distributed in proportion to mass and displacement at the discretized mass locations. Thus the design force at mass i is:

$$F_i = V_B (m_i \Delta_i) / \sum_{i=1}^{n} (m_i \Delta_i) \tag{33}$$

Similarity with force-based design will immediately be apparent. The difference is that the design inelastic displacement profile, rather than a height-proportional displacement is used. No additional force at the roof level, as is commonly adopted in force-based design, is generally necessary, though for tall flexible frames recent studies (Pettinger and Priestley, 2005) indicate improved performance if 10% of the base shear is allocated at roof level. When capacity design principles are adopted and normal gravity-load requirements for upper levels are enforced, it is rare to find excessive ductility demand developing at the upper floor levels of frames under inelastic time-history analyses. Capacity design will ensure that hinges do not develop in the upper levels of cantilever wall structures.

4.6 Analysis of Structure under Design Forces

Analysis of the structure under the lateral force vector represented by Equation (33) to determine the design moments at potential plastic hinge locations is straightforward, but needs a little careful consideration. In order to be compatible with the substitute structure concept that forms the basis of DDBD, member stiffnesses should be representative of effective secant stiffnesses at design displacement response.

For cantilever wall buildings, this can be simplified to simple distribution of the vertical force vector between walls in proportion to l_{w2}, as suggested above, with the walls then analysed separately.

For frame and dual (wall/frame) system buildings, more care is needed. With weak-beam/strong-column frame designs, beam members will be subjected to inelastic actions, and the appropriate stiffness will be:

$$I_{be} = I_{cr} / \mu_b \tag{34}$$

where I_{cr} is the cracked-section stiffness and μ_b is the expected beam displacement ductility demand. Analyses have shown that the member forces are not particularly sensitive to the level of stiffness assumed, and thus it is acceptable to assume that $\mu_b = \mu_s$, the frame design ductility.

Since the columns will be protected against inelastic action by capacity design procedures, their stiffness should be taken as I_{cr} with no reduction for ductility. Note that in the initial stages of the design, the beam and column strengths will not be known and it will thus not be possible to accurately define the cracked section stiffnesses of the beams or columns. However, as noted above the member forces are not strongly dependent on the stiffnesses, and it is the relative effective stiffnesses of beams and columns that are of importance, rather than the absolute values. Thus average values may be taken with adequate accuracy. Displacements predicted from the analyses may have significant errors, but since the displacement is in fact the design input, this is of little consequence.

A modification to the column stiffness must be made for the columns between ground and first floor. Plastic hinges will normally be expected at the base level, but not at first floor level. It has been found that the most appropriate way to model this in an elastic analysis is to place a hinge at the base level, and apply a base resisting moment Mb to the hinge, while representing the column by the cracked-section stiffness. This is illustrated in Figure 8(a). The values of moment placed at the base hinges are, to some extent, the designer's choice, since analysis of the structure under the calculated lateral force vector together with the chosen base moments M_b will ensure a statically admissible solution for the resulting design moments. In fact this freedom, implying some moment redistribution between beam hinges and column base hinges, allows the designer to improve the structural efficiency. A common choice will locate the point of contraflexure between 55% and 65% of the storey height above the base, thus ensuring capacity protection against hinging at the top of the ground floor columns, and an advantageous distribution of moments above and below the first level beams.

With a point of contraflexure chosen at 60% of the column height h_1 (to the beam centerline), and with reference to Figure 8 (a), equilibrium requires that

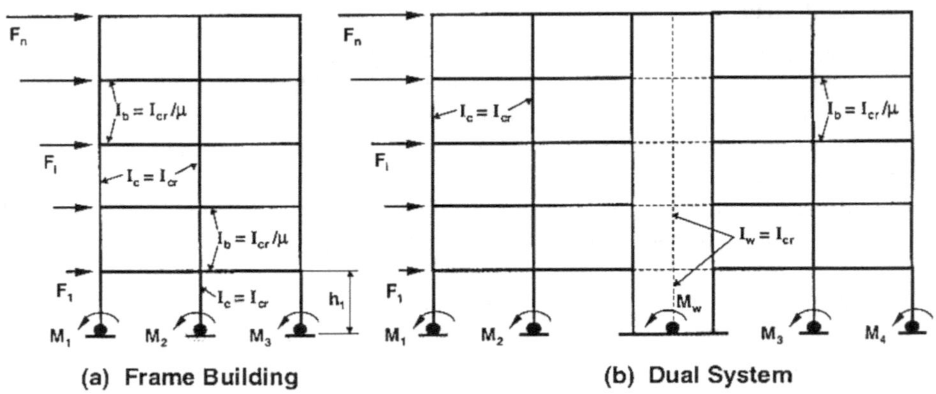

Figure 8. Member Stiffnesses for Analysis for Member Forces.

$$\sum M_b = M_1 + M_2 + M_3 = \sum_{i=1}^{n} F_i (0.6h_1) = V_b (0.6h_1) \qquad \textbf{(35)}$$

With dual structural systems consisting of walls and frames (Figure 8(b)), a similar approach is needed, with the frame represented by ductile beams of reduced stiffness, and columns with elastic stiffness. The wall stiffness will need to be reduced over the lower levels in proportion to the expected ductility demand. A decision will be require as to the proportion of total base shear force to be carried by the frame and by the wall. This will then dictate the base moments in the columns in a similar fashion to that represented by Equation (35) for pure frames. It has recently been shown (Paulay, 2002) that an attractive design decision is to allocate a constant shear force to the frames for the full height of the building, thus resulting in equal seismic demands on the beams for the full height.

Note that force-based designs using elastic analysis are found to have the wall dominating behaviour in the lower levels, and the frame dominating the upper storeys. A substitute structure analysis representing conditions at maximum displacement response is likely to result in significantly different distributions of actions between walls and frames.

4.7 Capacity Design for DDBD

Direct Displacement-based Design is a method for determining the optimum bending moments at intended plastic hinge locations, to satisfy a particular limit state. Having determined the design bending moments, normal capacity design procedures (Paulay and Priestley,1992) must be implemented to ensure that plastic hinges cannot develop at unintended locations, and also to ensure shear failure cannot occur. This requires a re-examination of existing capacity design procedures, particularly for wall and dual wall/frame structures. More definitive guide lines will shortly be available (Priestley et al., 2007).

5 Some Implications of DDBD

5.1 Influence of Seismic Intensity on Design Base Shear Strength.

Force-based and direct displacement-based design procedures imply significantly different structural sensitivity to seismic intensity. This can be illustrated with reference to Figure 9, where acceleration spectra (Figure 9(a)), and displacement spectra (Figure 9(b)) are shown for two seismic zones with different design intensities. It is assumed that the spectral shapes for the two zones are identical, and the design spectra in each zone are found by multiplying a base-level spectrum by the zone intensity factors Z_1 or Z_2.

We assume that structures are designed to satisfy the seismic design requirements for the two zones, and we further assume that the structural geometry, including member sizes (but not reinforcement contents) are identical for the two buildings. If the buildings are designed by conventional force-based procedures, the fundamental periods of the two buildings will be assumed to be the same. It is thus clear (see Figure 9(a)) that the required base-shear design forces V_{b1} and V_{b2} for the two buildings are related by:

$$V_{b2} = V_{b1}\frac{Z_2}{Z_1} \qquad \textbf{(36)}$$

Under direct displacement-based design, the assumption of equal geometry ensures that the yield displacements, and the limit-state design displacements for the two buildings are the same. Hence the ductility, and also the effective damping will also be the same for the two buildings. As may be seen from Figure 9(b), with equal design displacements and damping, the effective periods at design displacement response will be:

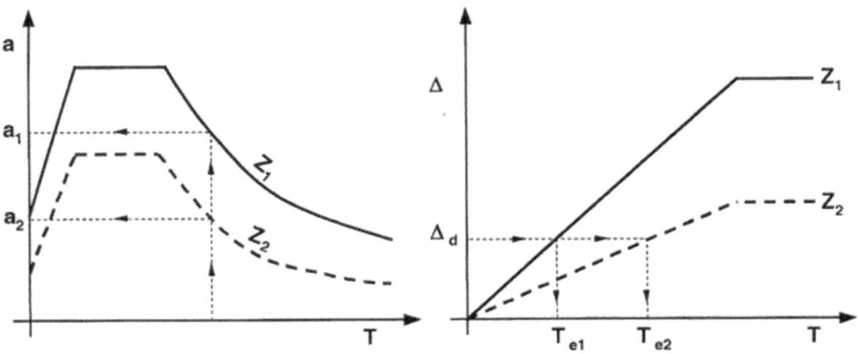

(a) Acceleration Spectra (b) Displacement Spectra

Figure 9. Influence of Seismic Intensity on Design Base Shear Force

related to the zone intensity by:

$$T_{e2} = T_{e1} \frac{Z_1}{Z_2} \quad \textbf{(37)}$$

From Equation (1) the effective stiffnesses are inversely proportional to the period squared, hence:

$$K_{e2} = K_{e1} \left(\frac{Z_2}{Z_1} \right)^2 \quad \textbf{(38)}$$

Further, since the design displacements are equal, Equation (2) yields the ratio of base shear forces as:

$$V_{B2} = V_{B1} \left(\frac{Z_2}{Z_1} \right)^2 \quad \textbf{(39)}$$

Thus the required base shear strength is proportional to the square of the seismic intensity. This is a fundamentally important difference between the two approaches, particularly for regions of low (or very high) seismicity.

5.2 Influence of Building Height on Required Frame Base Shear Strength

A further finding of some interest can be obtained by examining the sensitivity of required base shear strength of building frames to building height. We assume for simplicity that the section dimensions of structural members are not affected by building height, and that the design deflected shape is also independent of building height. Clearly this latter assumption will become increasingly crude when large variations in building height are considered. Let n = number of storeys, with constant mass m per storey. In the following, k_1 to k_6 are constants.

Effective mass: $$m_e = k_1 nm \tag{40a}$$

Design Displacement: $$\Delta_d = k_2 n \tag{40b}$$

With the above assumptions, both the yield displacement, and the design displacement will be proportional to height. Hence the design displacement ductility and thus the effective damping will be independent of height. Provided that the design displacement given by Equation (40b) is less than the displacement at peak period (e.g. $T_e = 4$ sec. in Figure 1(d)), the effective period will thus be:

$$T_e = k_3 \Delta_d = k_3 k_2 n \tag{41}$$

From Equation (2), the effective stiffness will be:

$$k_e = \frac{4\pi^2 m_e}{T_e^2} = \frac{k_4 k_1 nm}{(k_3 k_2)^2 n^2} = k_5 \frac{m}{n} \tag{42}$$

From Equations (1), (40a) and (40b), the design base shear will be

$$V_B = k_e \Delta_d = k_5 \frac{m}{n} k_2 n = k_6 m \tag{43}$$

Recalling that m is the mass of one storey, it is seen that the design base shear strength is independent of the number of storeys. This might seem to point the way towards further possible design simplifications.

6 Conclusions

Displacement-based seismic design characterizing the structure by structural properties applicable at the maximum displacement response resolves many of the problems implicit in current force-based seismic design where the structure is characterized by initial elastic stiffness (which is unknown until the design has been completed, and (typically) 5% damping. A more rational distribution of strength through the structure is obtained, and it becomes possible to design structures that achieve a specified limit state, based on acceptable damage criteria, under a specified level of seismic action. The procedure is simple, and intellectually satisfying.

7 References

Gulkan, P. and Sozen, M. (1974). Inelastic Response of Reinforced Concrete Structures to Earthquake Motions. *ACI Journal, 71(12), pp. 604-610*

Paulay, T. and Priestley, M.J.N. (1992). Seismic Design of Reinforced Concrete and Masonry Structures. *John Wiley and Sons, New York*

Paulay, T. (2002). A Displacement-Focused Seismic Design of Mixed Building Systems. *Earthquake Spectra, 18(4), pp. 689-718*

Priestley, M.J.N. (1993). Myths and Fallacies in Earthquake Engineering – Conflicts Between Design and Reality. *Bulletin, NZ National Society for Earthquake Engineering, 26(3), pp. 329-341*

Priestley, M.J.N. (2003). Myths and Fallacies in Earthquake Engineering, Revisited. The Mallet Milne Lecture. *IUSS Press, Pavia, Italy*

Priestley, M.J.N., Calvi, G.M. and Kowalsky, M.J. (2007). Direct Displacement-Based Seismic Design of Structures. *IUSS Press, Pavia (in press)*

Shibata, A. and Sozen, M. (1976). Substitute Structure Method for Seismic Design in Reinforced Concrete. *J. Structural Division, ASCE, 102(12), pp. 3548-3566*

Design and Assessment of Bridges *

Gian Michele Calvi[1], Nigel Priestley[2]

[1] Department of Structural Mechanics, University of Pavia, Pavia, Italy
[2] Emeritus Professor of Structural Engineering, University of California, San Diego, USA

Abstract. The most relevant improvement in the design methods for bridge structures traditionally followed new evidence in damage and collapse due to the major earthquakes. This logical process is critically review through the description of the most recent experience in bridge design and assessment, emphasizing its progressive evolution from acting force and strength, to ductility capacity and demand, to displacement – based approaches. Some controversial or often neglected aspects, related to seismic input, response and design issues are discussed.

1. Development of the Seismic Design Approaches

New structures were first routinely designed for earthquakes from 1930's. The design approach was borrowed from the same concepts regarding the design for wind forces: the structures were designed to remain in the elastic range for a constant fraction of the gravity weight, applied as a uniform lateral force. The main well known consequences (Asheim and Moehle, 1992, ASCE, 2000, ATC, 1997a and 1997b) related to the inadequacy of this elastic design approach, basically are: severe underestimations of the seismic deflections and inadequate combinations of action patterns produced by gravity and earthquake (due to the artificially low seismic forces). In particular, the latter deficiency may result in mislocating points of contraflexure, premature termination of reinforcement and neglecting of any detailing capable of favoring large inelastic deformation without significant strength degradation.

In the 1950's, as a consequence of the relevant damages and collapses, of the increased understanding of the dynamic characteristics of the seismic structural response and as it became realized that structures survived levels of response accelerations that apparently exceeded those corresponding to the ultimate strength, the concept of "ductility" was adopted. Historically, engineers have always been more comfortable designing for "loads" than for deformation-inducing actions. This new concept, in fact, substantially was an attempt to reconcile inconsistencies in the fundamental basis of force-based design, attributing to the capacity of a structure to deform inelastically without significant strength loss, the reason for surviving an earthquake that would have required more strength than that available to respond elastically. A considerable amount of research was developed in order to correlate the concept of ductility to a sort of equivalent strength, which represents the seismic resistance of the structure. Consequently, the well-known concepts of conservation of acceleration, velocity and displacement as a function of the fundamental period of vibration of the structure was introduced and developed.

In the present time, the majority of the codes of practice are still based on these concepts. Essentially, elastic acceleration spectra are considered as a function of an assumed ductility capacity of bridges, and capacity design principles are applied to assure that the assumed post elastic mechanism will develop, avoiding potential brittle damage modes. In spite of the great development of a large number of new design approaches that require increased emphasis on displacement, the most common approach has been to attempt to modify force-based design procedures, rather than a complete, more rational, revision of the seismic design procedure.

2. Fundamental Aspects Related to the Seismic Design of Bridges

Although the force-based approach is the traditional reference method included in the majority of the seismic design codes, it is characterized by several common fallacies, which will be discussed in the following of this section, in order to highlight the effectiveness of a displacement-based approach, particularly for long period structures.

Probably the most rational displacement-based approach (direct displacement base design, DDBD) will be schematically presented, since it is fully illustrated in another lecture note.

Afterwards, two particularly important issues related to the seismic design of bridges, such as the determination of the elastic stiffness and the ductility capacity of bridge piers, will be discussed in detail, in order to better understand their role and their evaluation within the seismic design.

2.1. Common Fallacies

Despite bridges have been designed by reference to acceleration response spectra for the past 40 years, the assumption that an elastic acceleration spectrum provided the best means for assessing the seismic response of a structure has been proven to be a fallacy. In this section, the most important controversial issues concerning the design process are briefly summarized.

Firstly, only "snapshots" of transient behaviour at maximum modal response are considered, ignoring the real duration effects of the earthquake. Moreover, the combination rules of these modal maxima are of dubious relevance for inelastic structural response and it is implicitly assumed that the maximum transient response is more relevant than the final configuration of the structure after the earthquake, which is not considered in the design process. Actually, the role played by the residual deformations is so important that they can be related to a damage index to quantify the performance level of a structure under seismic loading.

Usually, the level of damage is related to material strains, which in turn can be related to the maximum response displacements (but not to response accelerations). For this reason, the importance of estimating peak displacement response, evaluated quite differently according to different approaches and codes, should be adequately emphasize. Generally the estimation of the design displacement involves modification of the displacements of the corresponding elastic system of equal initial stiffness and unlimited strength, assuming the following equal displacement approximation:

$$\Delta_{\max,duct} = \Delta_{\max,elastic} = \frac{T^2}{4\pi^2} \cdot a_{(T)} g \qquad \textbf{Eq. 1}$$

Where $\Delta_{max,duct}$ and $\Delta_{max,elastic}$ are the equal maximum displacement related to the ductile and to the elatic response spectrum respectively. T is the period of the structure and $a_{(T)}g$ the corresponding pseudo-spectral acceleration. It is well known that the equal displacement approximation is non-conservative for short-period structures. If the equal energy approach is applied, equating the energy absorbed by the inelastic system, on a monotonic displacement to peak response, to the energy absorbed by the equivalent elastic system with same initial stiffness, the peak displacement of the inelastic system is:

$$\Delta_{\max,duct} = \Delta_{\max,elastic} \cdot \left(\frac{R^2+1}{2R}\right) = \frac{T^2}{4\pi^2} \cdot a_{(T)} g \cdot \left(\frac{R^2+1}{2R}\right) \qquad \textbf{Eq. 2}$$

Where R is the design force reduction factor; in accordance with the UBC (1997), the maximum displacement related to the ductile spectrum is estimated as: $\Delta_{max,duct} = 3\Delta_y\ R\ /\ 8$, hence its relation with the maximum elastic displacement is: $\Delta_{max,duct} = 3\ \Delta_{max,elastic}\ /\ 8$.

Priestley (2003) showed that all formulations are correct at some part of the period range of structural response, and all are wrong at other periods. This was explained with reference to typical displacement spectra, reproduced in Figure 1, where the influence of inelastic response was represented by a lengthening of the effective period of response, with hysteretic damping being represented as equivalent viscous damping. For short period structures, the increase in displacement response from period elongation is less than the decrease resulting from increased damping. For medium period structures, the two effects almost balance each other. For long period structures the period elongation does not result in significant displacement increase, and the influence of increased damping is to reduce the overall displacement response. For very long period structures, the displacement is equal to the ground displacement, independent of period and damping.

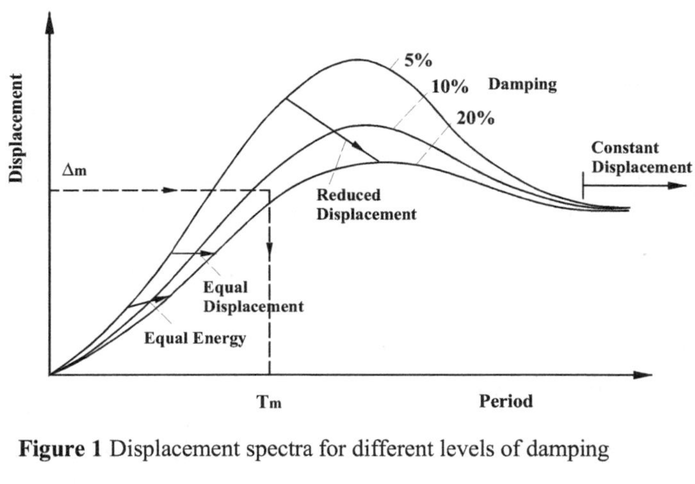

Figure 1 Displacement spectra for different levels of damping

It was also noted (Priestley, 2003) that the elastic acceleration approach placed excessive emphasis on the elastic stiffness characteristics of the structure and its elements. For reinforced concrete and masonry structures the estimation of these stiffness values varies greatly between different design codes. Further, these elastic characteristics only pertain to low level seismic response, and are permanently modified as soon as the structure exceeds yield.

Other issues of fundamental importance concern:

1. Interdependency of strength and stiffness, meaning that stiffness (and hence natural periods, elastic strengths, and strength distribution through the structure) cannot be accurately determined until the structure is fully designed;
2. Inadequate representation of variations of hysteretic characteristics of different structural systems;
3. Simplistic and inappropriate definition of behaviour factors for whole categories of structures, and a lack of appreciation that ductility capacity can vary widely within a structural class – this results has been explicitly described for the case of bridges in Pinto et al. (2003), where it was shown that a standard design procedure based on a force reduction factor approach cannot guarantee a uniform level of protection;
4. Inadequate representation of the influence of foundation flexibility on seismic response;
5. Inadequate representation of structural performance of systems where inelastic action develops in different members at different levels of structural response (e.g. bridges with columns of different heights, marginal wharves with ductile piles of different heights, structural wall buildings with walls of different lengths);
6. Inadequate representation of structures with dual load paths (e.g. a bridge with an elastic load path involving superstructure action spanning between abutments, and an inelastic load path involving ductile action of the piers).

In addition to the issues previously described, in the case of seismic design of bridges, and in general for structure characterized by long periods of vibration, it has also to be highlighted that additional concerns may arise. For instance, in case of long period structures, acceleration spectra are characterized by relatively low levels of acceleration and design may be based on minimum levels rather than by a proper estimate of the demand. The possible presence of strong pulses, typical of near fault action may have a moderate impact on the spectral shape (Somerville, 2003). On the opposite, it has to be recognized that large displacement demand are typical of long period structures, and near fault effects have a tremendous impact on this parameter. This evidence points towards the use of displacement spectra as a better way to represent the design action.

2.2. Fundamentals of the DDBD Method

A number of displacement-based (sometimes referred to as performance-based) design methods were developed (*fib*, 2003) in the last years. In general, and more specifically in one of the approaches (Priestley and Calvi, 2003), the fundamental difference from force-based design is that the structure to be designed is characterized by a single-degree-of-freedom (SDOF) representation of performance at peak displacement response, rather than by its initial elastic characteristics. The method is based on the Substitute Structure Approach (Gulkan and Sozen, 1974, Shibata and Sozen, 1976). The approach attempts to design a structure which would achieve, rather than be bounded by, a given performance limit state under a given seismic intensity, essentially resulting in uniform-risk structures, which is philosophically compatible with the uniform-risk seismic spectra incorporated in most design codes. The design procedure determines the strength required at designated plastic hinge locations to achieve the design objectives in terms of defined displacement objectives. A capacity design procedures has then to be applied, in order to ensure that plastic hinges occur only where intended, and that non-ductile modes of inelastic deformation do not develop.

Among the large number of alternative seismic design philosophies proposed in the last years (*fib*, 2003), the procedure known as Direct Displacement-Based Design (DDBD) (Priestley and Calvi, 2003) deserve a particular mention. This design method appears to be more intellectually satisfying than the alternatives, it is the best equipped to address the deficiencies of conventional force-based design, simple to apply and better suited to incorporation in design codes. Since this design method is fully illustrated and well commented in another lecture note, in this section only a schematic overview is presented.

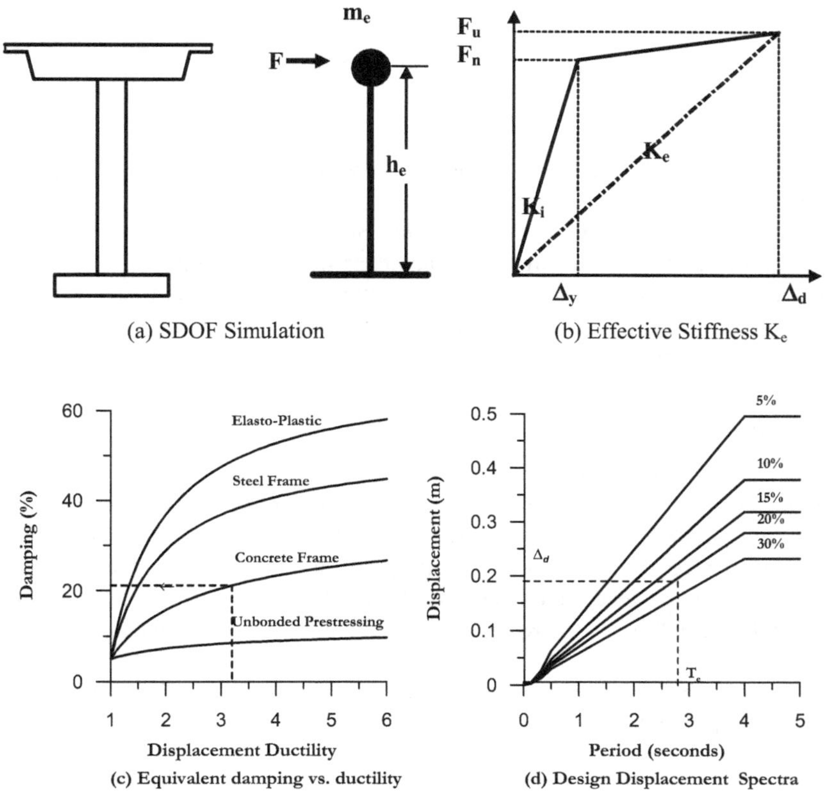

Figure 2 Fundamentals of the Direct Displacement Based Design.

The method can be described with reference to a SDOF model of a bridge column (Figure 2, though the basic fundamentals apply to all structural types) and its bilinear envelope of the lateral force-displacement response, characterized by an initial elastic stiffness K_i, followed by a post yield stiffness rK_i,. While force-based seismic design characterizes a structure in terms of elastic, pre-yield, properties (initial stiffness K_i and elastic damping), DDBD characterizes the structure by secant stiffness K_e at maximum displacement Δ_d and a level of equivalent viscous damping ξ_e, representative of the combined elastic damping and the hysteretic energy absorbed during elastic response. With the design displacement at maximum response determined, and the corresponding damping estimated from the expected ductility demand, the effective period T_e at maximum displacement response can be read from a set of displacement spectra for different levels of damping. The effective stiffness K_e of the equivalent SDOF system at maximum displacement and the design base shear force can be found respectively by the following Eq. 3 and Eq. 4:

$$K_e = 4\pi^2 m_e / T_e^2 \qquad \textbf{Eq. 3}$$

$$F = V_B = K_e \Delta_d \qquad \textbf{Eq. 4}$$

where m_e is the effective mass of the structure participating in the fundamental mode of vibration.

The design concept is thus very simple. Such complexity that exists relates to the determination of the “substitute structure” characteristics, of the design displacement and of the development of design displacement spectra. Careful consideration is however necessary for the distribution of the design base shear force V_b throughout the structure and the analysis of the structure under the distributed seismic force.

2.3. Elastic Stiffness

In force-based design, the elastic stiffness is required at the start of the design, in order that the elastic periods of the structure can be defined, and also at a later stage of the design to distribute the total design inertia force to members in proportion to their initial stiffness. It is only recently that it has been recognized that it is inappropriate to use the uncracked stiffness in period calculations. Now, some recognition is made of the reduction in stiffness caused by cracking, and it is common to use 50% of the gross section moment of inertia in estimating section stiffnesses.

More realistically, stiffness can be assessed from the moment-curvature relationship for a section in accordance with the beam equation:

$$EI_{eff} = M_N / \phi_y \qquad \textbf{Eq. 5}$$

where M_N is the nominal moment capacity of the section, and ϕ_y is the yield curvature of the equivalent bilinear representation of the moment-curvature curve. This can be explained with reference to Figure 3, which shows a typical moment-curvature relationship together with a bilinear approximation for a 2 m diameter bridge column with 2% longitudinal reinforcement ratio, and light axial load. It is accepted by the research community that the most appropriate linearization of moment-curvature relationships is the following: an initial elastic segment passing through the point identifying the "first yield" and extrapolated to the nominal flexural strength M_N; is combined with a post-yield segment connected to the ultimate strength and curvature. "First yield" of the section is defined as the moment M_y and curvature ϕ'_y when the section first attains the reinforcement tensile yield strain of $\varepsilon_y = f_y/E_s$, or the concrete extreme compression fiber attains a strain of 0.002, whichever occurs first. The nominal flexural strength M_N develops when the extreme compression fiber strain reaches 0.004, or the reinforcement tension strain reaches 0.015, whichever occurs first. Thus, the yield curvature is given by:

$$\phi_y = \phi'_y \, M_N / M_y \qquad \textbf{Eq. 6}$$

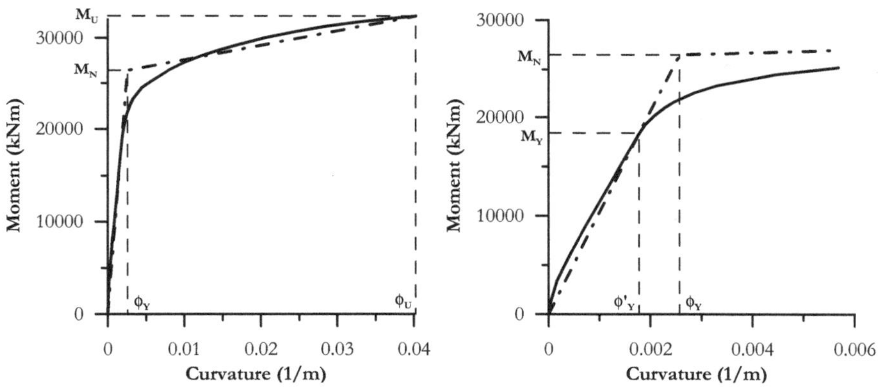

Figure 3 Typical column moment – curvature relationship

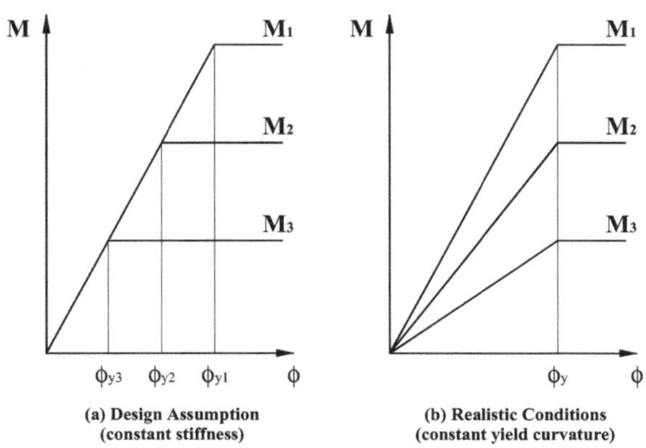

Figure 4 Influence of Flexural Strength on Moment-Curvature Relationship

Experimental evidence, and detailed analytical results, indicates that the common assumption of stiffness being independent of strength is not valid. Actually, yield curvature is effectively independent of strength, and hence the stiffness is directly proportional to the flexural strength, as is seen from Eq. 5 with ϕ_y a constant. The correct relationship is thus illustrated in Figure 4.

An example of moment-curvature curves computed on circular bridge piers is shown in the following figure for two levels of flexural reinforcement ratio and a range of axial load ratios. Only the initial part of the moment-curvature curves has been included, to enable the region up to, and immediately after yield to be clearly differentiated. Also shown in the figure are the calculated bi-linear approximations for each of the curves. Note that the apparent over-estimation by the bi-linear representations of the actual curves is a function of the restricted range of curvature plotted.

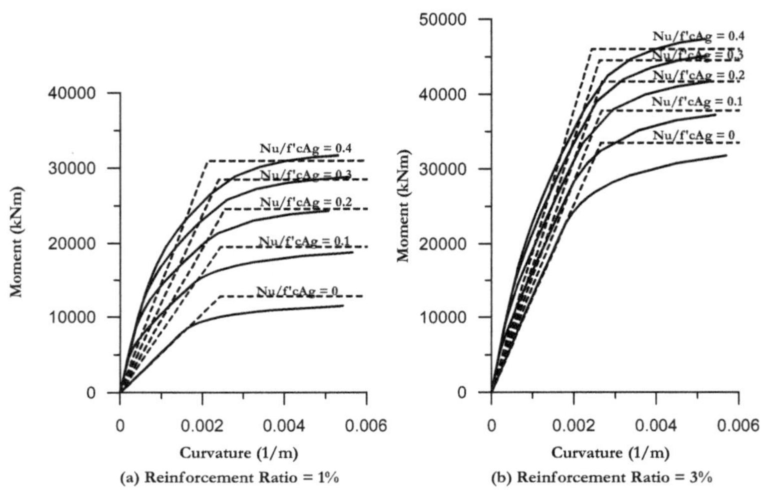

Figure 5 Selected Moment Curves for Circular Concrete Bridge Columns

The nominal moment capacity and the equivalent bi-linear yield curvature are plotted in dimensionless form in Figure 6. The influence of both axial load ratio and reinforcement ratio on the nominal moment capacity is, as expected, substantial with a range between maximum and minimum values. On the other hand, it is seen that the dimensionless yield curvature is comparatively insensitive to variations in axial load or reinforcement ratio. Thus the yield curvature is insensitive to the moment capacity. The average value of dimensionless curvature of

$$\phi_{Dy} = \phi_y D / \varepsilon_y = 2.25 \qquad \textbf{Eq. 7}$$

is plotted on Figure 6b, together with lines at 10% above and 10% below the average. It is seen that all data with the exception of those for low reinforcement ratio coupled with very high axial load ratio fall within the +/–10% limits.

It should be noted that though the data were generated from a specific column size and material strengths, the dimensionless results can be expected to apply, with only insignificant errors, to other column sizes and material strengths within the normal range expected for standard design. The results would not, however, apply to very high material strengths (f'_c > 50MPa, or f_y > 600MPa) due to variations in stress-strain characteristics.

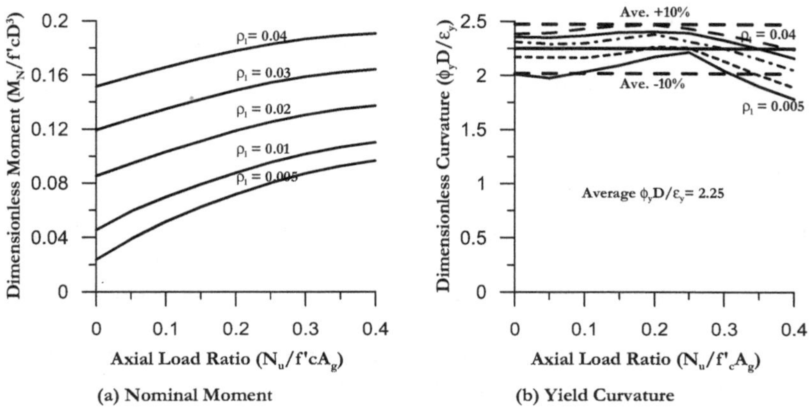

Figure 6 Dimensionless nominal moment and yield curvature for concrete bridge columns

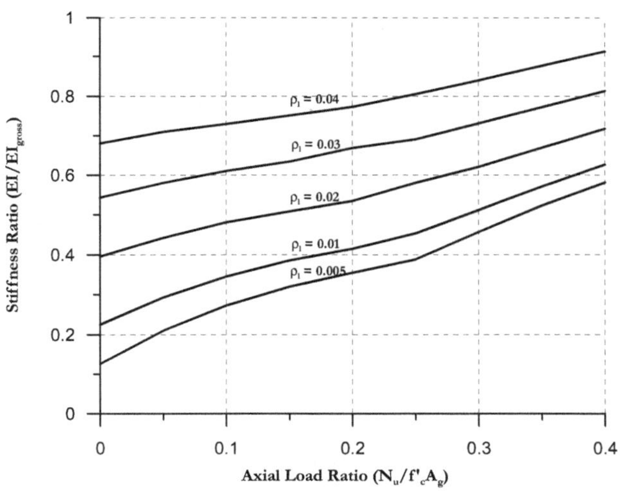

Figure 7 Effective Stiffness of Circular Bridge Columns

The data in the previous Figure 5 and Figure 6 can be used to determine the effective stiffness of the columns as a function of axial load ratio and reinforcement ratio. The ratio of effective stiffness to initial uncracked section stiffness is thus given by:

$$EI_{eff} / EI_{gross} = M_N /(\phi_y EI_{gross})$$ **Eq. 8**

Results are shown in Figure 7. It will be seen that the effective stiffness ratio varies between 0.13 and 0.91. Clearly the assumption that stiffness is a constant for a given section, regardless of strength, is unacceptable.

2.4. Ductility Capacity of Bridge Columns and Elastic Flexibility of Capacity-Protected Members

A common assumption of force-based design is that structures of a particular material and class (e.g. concrete; bridge column) have a constant ductility capacity. This is reflected in the specification of a constant force-reduction factor for all structures in that class. Simple evaluation of the basic equations defining ductility capacity reveals that this is inappropriate. An example of the influence of structural geometry on displacement capacity is provided in Figure 8, which compares the ductility capacity of two bridge columns with identical cross-sections, axial loads and reinforcement details, but with different heights. The two columns have the same yield curvatures ϕ_y and ultimate curvatures ϕ_u and hence the same curvature ductility factor $\mu_\phi = \phi_u/\phi_y$. Yield displacements, however, may be approximated by:

$$\Delta_y = \phi_y H^2 / 3$$ **Eq. 9**

where H is the effective height, and the plastic displacement $\Delta_p = \Delta_u - \Delta_y$ by:

$$\Delta_P = \phi_P L_P H$$ **Eq. 10**

where $\phi_p = \phi_u - \phi_y$ is the plastic curvature capacity. The displacement ductility capacity is thus given by:

$$\mu_\Delta = \frac{\Delta_y + \Delta_P}{\Delta_y} = 1 + 3\frac{\phi_p L_P}{\phi_y H}$$ **Eq. 11**

where L_p is the plastic hinge length.
For circular bridge columns, the plastic hinge length can be expressed as:

$$L_P = 0.08 \cdot H + 0.022 \cdot f_y \cdot d_b$$ **Eq. 12**

where f_y and d_b are respectively the yield stress (in MPa) and the diameter of the flexural reinforcement in the plastic hinge region. Using this approach, it is found that the squat column of Figure 8a has a displacement ductility capacity of $\mu_\Delta = 9.4$, while for the more slender column of

Figure 8b, μ_Δ= 5.1. Clearly the concept of uniform displacement ductility capacity, and hence of a constant force-reduction factor is inappropriate for even this very simple class of structure.

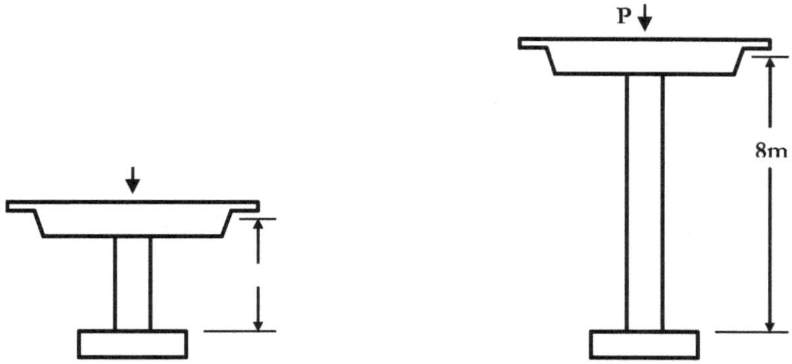

(a) Squat Column, $\mu_\Delta = 9.4$ (b) Slender Column, $\mu_\Delta = 5.1$

Figure 8 Influence of height on displacement ductility capacity of circular columns ($P = 0.1f'_cA_g$; 2% longitudinal, 0.6% transverse reinforcement)

It can easily be shown (Priestley, 2003) that the elastic flexibility of the capacity-protected members influences the displacement ductility capacity of the structure, and hence might be expected to influence the choice of force-reduction factor in force-based design. Consider the simple two-column bridge bent illustrated in the next Figure 9: the column bases are connected to the footings by pinned connections, and thus no moments can develop at the base. Plastic hinges are intended to form only at the top of the columns.

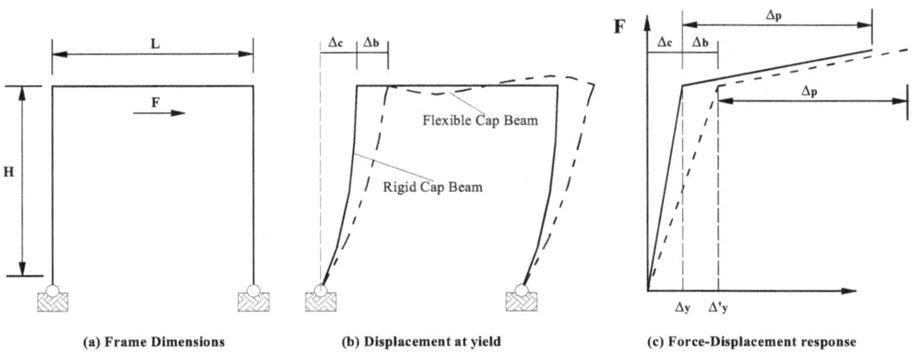

Figure 9 Influence of cap-beam flexibility on ductility capacity of two-column bridge bent

Consider first the case where the cap beam is assumed to be rigid. The yield displacement under lateral forces F is thus $\Delta_y = \Delta_c$ resulting solely from column flexibility. All plastic displacement originates in the column plastic hinge regions, since the design philosophy requires the cap beam to remain elastic. With a plastic displacement of Δ_p corresponding to the rotational capacity of the column hinges, the structure displacement ductility is:

$$\mu_{\Delta r} = 1 + \frac{\Delta_p}{\Delta_c} \qquad \textbf{Eq. 13}$$

Cap beam flexibility will increase the yield displacement to $\Delta_y = \Delta_c + \Delta_b$, where Δ_b is the additional lateral displacement due to cap beam flexibility, but will not result in additional plastic displacement, since this is still provided solely by column hinge rotation. For bent dimensions H x L and cracked-section moments of inertia for beam and columns of I_b and I_c, respectively, the yield displacement is now:

$$\Delta_y = \Delta_c + \Delta_b = \Delta_c\left(1 + \frac{0.5 I_c L}{I_b H}\right) \qquad \textbf{Eq. 14}$$

and the structural displacement ductility capacity is reduced to:

$$\mu_{\Delta f} = 1 + \frac{\mu_{\Delta r} - 1}{1 + 0.5 I_c L / I_b H} \qquad \textbf{Eq. 15}$$

As an example, take $L = 2H$, $I_b = I_c$, and $\mu_{\Delta r} = 5$. From Eq. 15 it is found that the displacement ductility capacity is reduced to $\mu_{\Delta f} = 3$. Again it would seem to be inappropriate to use the same force reduction for the two cases. This effect is not included in any design codes, and is rarely adopted in force-based design practice.

3. Fundamental Aspects Related to the Seismic Input and Effects on the Response of Isolated Bridges

3.1. Displacement Spectra

It is well known (Priestley and Calvi, 2003) that design displacement spectra for a given level of damping can be generated from the acceleration spectra using the approximate relationship:

$$\Delta_{T,5} = \frac{T^2}{4\pi^2} \cdot a_{T,5}\, g \qquad \textbf{Eq. 16}$$

where $\Delta_{T,5}$ and $a_{T,5}$ are the response displacement and acceleration coefficient for period T and 5% damping.

Recent work (Bommer et al., 2000, Faccioli et al., in preparation) has enabled (Priestley and Calvi, 2003) to generate approximate elastic displacement spectra for strike-slip earthquakes of different magnitudes. The spectra are characterized by a linear increase in response displacement with increasing period, followed by a plateau. At high periods the response displacement eventually decreases to the peak ground displacement, but this will rarely be of significance for bridge design (except, perhaps, suspension and cable-stayed bridges).

The corner period T_C between the linearly increasing and plateau part of the spectrum, and the maximum displacement δ_{max} for firm ground can be related to the moment magnitude M_W by the following expressions:

$$T_c = 1.0 + 2.5(M_w - 5.7) \quad \textbf{Eq. 17}$$

$$\delta_{\max} = \frac{10^{(M_W - 3.2)}}{r} \text{ (mm)} \quad \textbf{Eq. 18}$$

where r is the nearest distance from the site to the fault plane in *km*. For rock and soft soil sites, the displacement given by the previous equation should be multiplied by 0.7 and 1.5, respectively, to give approximate representation of soil conditions (CEN, 1998).

Eq. 17 and Eq. 18 are plotted in figure 3 for different earthquake magnitudes and for different distances from the fault plane. It should be noted that these apply for design spectra dominated by a single causative fault. The consequences in terms of uniform-risk spectral displacement shapes are less obvious than for specific earthquake intensities.

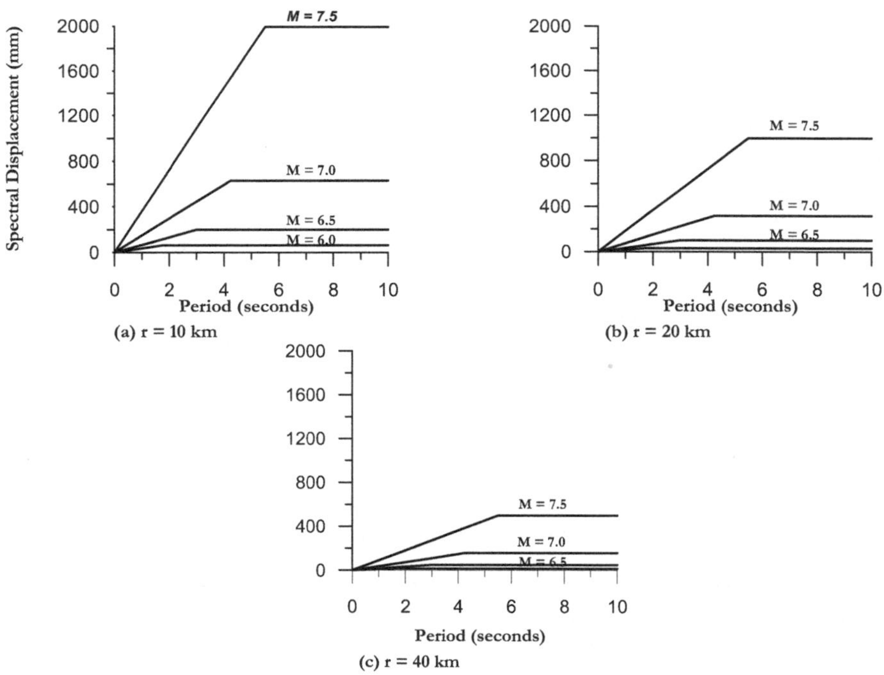

Figure 10 Influence of cap-beam flexibility on ductility capacity of two-column bridge bent

3.2. Vertical Component and Axial Force Variation and its Implications on Isolated Bridges

The importance of the vertical seismic input in the structural response of bridges has been investigated by various authors (Ambraseys and Douglas, 2000, Ghobarah and Elnashai, 1998, Elnashai and Papazoglou, 1997). The damaging effects of the vertical component are more evident in the near field since the vertical motion attenuates faster than the horizontal one: in this case ground motions from large earthquakes ($M_s > 7$) can produce significant horizontal and vertical components and the ratio of the vertical to horizontal maximum *PGA* may exceed 1.

The energy content of the vertical component is concentrated in a narrow high frequency range. This results in a possibly dangerous match with the vertical periods of common bridges, usually belonging to the higher frequency range. Furthermore, several records indicate that the maximum vertical response occurs 1 or 2 seconds earlier than transverse one, others show a coincidence in time. A compendium of field observations and analytical results indicates that certain failure modes are convincingly attributable to high vertical earthquake-induced forces, which, in addition to the possible overstressing in compression or tension, may induce shear or flexure failure. An increase in shear demand may be due to second order effects caused by the presence of high vertical dynamic forces.

The effects of axial force variations may be emphasized in presence of isolation systems. The case of friction pendulum systems (FPS, used for the upgrading of the Bolu viaduct, as discussed in the following chapter 4) offers an interesting example to discuss possible effects and consequences. FPS use geometry and gravity to achieve the desired isolated response, based on well-known principles of pendulum motion. Though friction pendulum systems may be strongly influenced by the axial load level acting at a given time, since a variation of the axial force results in corresponding variation of equivalent yielding level and in the post-yielding stiffness in the non-linear phase of the hysteretic response, and this, in turn may produce increments in the shear force demand and potential torsional effects on the piers.

The axial force variation on an isolation device is not only affected by the vertical acceleration, but also depends on a combination of effects due to horizontal input (because of the necessary dynamic equilibrium to the horizontal forces) and to the geometrical configuration (plan and elevation irregularities).

An extensive, though not conclusive, study on the subject (Calvi et al., 2004), based on parametric non-linear time history analyses, showed that the inclusion of axial force effects may not be significant for what concerns variation of the displacement demand, but may induce important increment of shear, bending and torsional moment demand on the piers.

The fundamental parameters that may amplify, or reduce, these effects are the ratio between deck and pier mass, the aspect ratio of the deck, the radius of curvature of the bridge, the intensity of the ground motion and the consideration of vertical input, as briefly discussed below.

a. Ratio between deck and pier mass: a significant variation of the shear force transmitted from the deck to the pier may result in strongly attenuated effect at the pier base when the ratio of the pier mass to the deck mass is high.

b. Aspect ratio of the deck: for the same level of horizontal force, the axial force variation possibly induced by the horizontal acceleration is higher for a deck section relatively larger and for devices relatively closer one to each other.

c. Radius of curvature of the viaduct: it is shown that a curved bridge may result in higher effects, due to the interaction of vertical and horizontal response.

d. Intensity of the ground motion: relatively high horizontal peak ground accelerations may induce more significant effects.

e. Consideration of vertical input: the inclusion of the vertical component of the input ground motion may result in being the crucial point to verify whether important effects have to be expected and considered.

These considerations may be of some help in deciding whether axial force effects may be neglected or should be considered in the analysis.

A fundamental aspect related to design concept should also be noted. Actually, when dealing with isolated bridges, it is common practice to assume that possible pier collapses are capacity protected by the shear capacity of the isolation system. This implies that there is no reason to protect a possibly brittle shear collapse mode with a lower strength flexural yielding of the pier. Clearly, this situation may not apply if a significantly higher shear force is transmitted from the deck to the pier. As a consequence, it is felt appropriate to recommend that when using friction pendulum systems capacity design principles are still applied to protect undesired failure modes of the pier and foundation system.

3.3. Near Fault Effects

Ground motion records from recent earthquakes confirmed that near fault ground motions are dominated by a large long period pulse of motion that occurs at on the horizontal component perpendicular to the strike of the fault. Preliminary equations to define appropriate response spectra, taking into account magnitude dependence of the pulse and rupture directivity effects have been proposed (Somerville, 2003), but it appears that significant developments are needed before transferring these scientific findings into common practice; permanent differential displacements may combine with strong pulses and directivity effects, resulting in anomalous demand.

With reference to displacement-based design approaches, it has been noted that the presence of a dominating single pulse in the ground motion results in a strongly diminished capacity of dissipating energy for the structure, and this may in turn results in larger displacement demand. Based on these considerations, on the equation recommended in EC8 (1998) to take into account different level of viscous damping and on the recognized equivalence between viscous damping and hysteretic energy dissipation, a modified correction factor has been proposed to evaluate the displacement demand in near fault conditions (Priestley, 2003). This proposal, described below, is based on empirical evidence and certainly needs validation and refinements, but has connotation of simplicity and applicability.

For normal accelerograms EC8 proposes the following expression to compute the displacements corresponding to a different level of damping, as a function of the displacement for 5% damping:

$$\Delta_{T,\xi} = \Delta_{T,5} \cdot \left(\frac{10}{5+\xi} \right)^{0.5} \qquad \textbf{Eq. 19}$$

In the near-field region, the influence of velocity pulses may reduce the effectiveness of damping (and hysteretic energy absorption), therefore Eq. 19 is likely to be non-conservative, and may be replaced by the following one:

$$\Delta_{T,\xi} = \Delta_{T,5} \cdot \left(\frac{10}{5+\xi} \right)^{0.25} \qquad \textbf{Eq. 20}$$

The difference of the two previous equation can be appreciated in the next Figure 11, for different levels of damping. In the case of the velocity-pulse type, the decrease of displacement due to an increase of damping is much less pronounced than the other case. As a consequence, given the design displacement and the damping, the associated period tends to decrease.

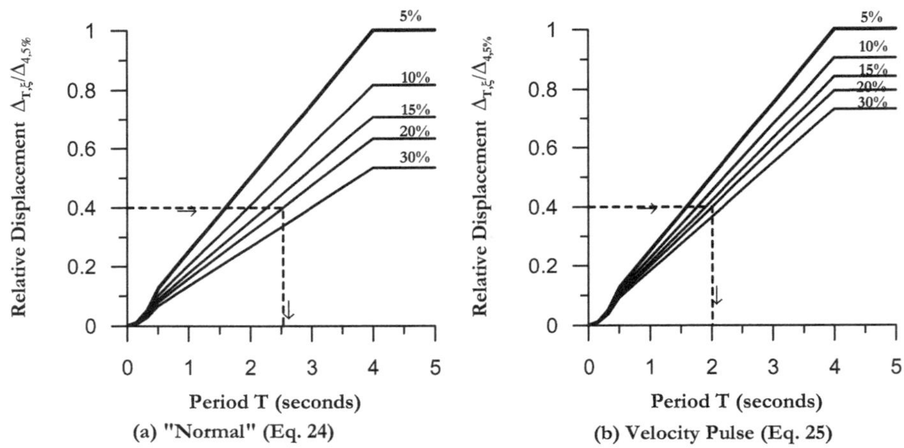

Figure 11 Influence of Damping on Displacement Response

3.4. Non-Synchronous Input

The potential relevance of the spatial variability of the seismic ground motion on the response of bridge structures has been recognized since a long time; however, a relatively little amount of research studies have been developed on the subject. It is accepted that the spatial variability results from the combination of loss of coherence due to wave propagation, time delay between arrival of wave trains and local filtering due to local site conditions. In a recent comprehensive study (Pinto et al., 2003) it has been pointed out that the accurate determination of the effects of these sources of differential ground motion requires data that are normally beyond the extension of potential studies in real design situations. In this study it has been shown that the ductility demand at the base of the piers increases considerably when the spatial variability is included, even for rather ordinary bridge geometries. As a consequence, it is shown that the probability of collapse may vary of more than one order of magnitude if these phenomena are considered.

In the mentioned paper a procedure to reduce the risk of collapse in presence of potential non-synchronous input is proposed. However, the limited extension of the study, the relatively poor amount of available data and the variability of the results obtained by different researchers (Sextos, 2003) and designers in practical cases (Teyssandier, 2003, Pecker, 2003, Priestley and Calvi, 2002) clearly indicate the need of research and code–oriented studies on this subject.

4. Assessment of Bridges and Recent Experience

4.1. Assessment and Strengthening of the Bolu Viaduct

Bolu Viaduct 1 (Figure 12), part of the Great Anatolian Highway, consists of two parallel bridges, each carrying a separate traffic direction, with the Ankara-bound bridge having 58 spans, and the Istanbul-bound bridge having 59 spans. Each span is approximately 39.4 m long, and is constructed from 7 precast V-girders with an in – situ topping. Each span was supported on pot-bearings designed for 200 mm maximum travel. A 1.5 m long link-span extension of the deck slab connected the spans into 10-span segments with movement joints between the 10-span segments. However, because of the bearing support detail, the gap between the girder ends, and the flexibility of the link-slab, the spans remained effectively simply supported for live loads as well as dead loads.

Figure 12 General view of the Bolu Viaduct (Priestley and Calvi, 2002)

At internal supports, the bridge is supported by tall hollow reinforced concrete piers of approximately rectangular section modified by architectural detailing, generally in the range 40 – 50 m high, though a number of shorter piers exist, particularly near the Istanbul abutment.

Piers are founded on massive reinforced concrete pile caps, in turn supported on twelve 1.8 m diameter Cast-In-Drilled-Hole (CIDH) piles passing through superficial soils of variable strength and bearing on alluvium layers, generally at about 30 m depth.

At the bridge ends, the V-girders were individually supported by pot-bearings on a seat-type abutment, again supported on CIDH piles, but of reduced number and size.

Seismic resistance relied primarily on a seismic isolation system consisting of "crescent-moon" steel energy dissipating units (EDU's) located at each support connecting the spans to a centrally mounted dissipator support block. At movement joints and the central pier of each 10-span segment, the EDU's incorporated sliders and lock-up pistons to allow relative thermal movements to occur freely, but to ensure full engagement of all EDU's under seismic loading. Displacement capacity of the EDU's was 480 mm. Transverse displacements were restrained by shear blocks adjacent to beams 3 and 5 as a back-up in the event of extreme displacements, and longitudinal relative movements at expansion joints were constrained by cable restrainers.

Description of the damage. On November 12, 1999 an earthquake of moment magnitude 7.2 occurred on the Duzce fault causing severe damage to tunnels and bridges under construction on the Great Anatolian Highway (Mitchell, 2000). Peak ground accelerations of the order of 0.8 g were estimated at the viaduct site, based on accelerograms recorded nearby. More important to the bridge performance, right – lateral fault slip of approximately 1.6 m occurred on a fault scarp traversing the bridge alignment, at an acute angle (approximately 15 degrees to the bridge longitudinal axis), resulting in shortening of the bridge length by about 1.5 m, concentrated over two spans of the bridge.

Displacements resulting from the fault slip and the vibratory response exceeded the capacity of the seismic isolation system. As a consequence, the EDU's were destroyed, and the pot-bearings at the beam ends were ejected. Impact between the ends of the central V-beam and the EDU support block occurred at most spans, destroying many of the support blocks and damaging many of the beam 4 ends. In the majority of the cases, the damage to beam ends was superficial, though in one case the damage included crushing of concrete and fracture of reinforcement for a distance of up to 4 m from the beam support.

Impact between the transverse shear restraint blocks and beams 3 and 5 caused extensive damage to the shear blocks, and some damage to beam ends, though this was superficial. Further damage to beam ends, applicable to all beams across the section, occurred as a consequence of unseating from the pot-bearings, with impact between the beam end and the bearing support block, or the pier cap. This damage is minor in all cases.

Consequently to the fault movement and the failure of the EDU's, residual displacement of the beam ends was considerable, being as high as 1100 mm longitudinally, and 500 mm transversely. In a number of cases this displacement was such that the beam ends remained unsupported, hanging over the edge of the pier cap. In such cases, and where the beams ends were unsupported (having moved beyond the edge of the bearing blocks) but were still within the plan area of the pier cap, support for the beam end was only provided by flexure of the link span joining adjacent spans. The factor of safety against failure in these cases, using conventional flexural strength theory, is less than 1.0, since the shear corresponding to flexural strength of the link slab is only 90% of that needed to support the reaction of the span dead-load, using expected (rather than nominal) material properties, and ignoring strength reduction factors. It will also be noted that during unseating of the beam ends, the shear in the link spans will have been increased by dynamic impacts, to perhaps twice the static value. It is apparent that catastrophic failure has been averted by large vertical relative displacements, up to 300 mm, which have occurred across these 1800 mm long link spans, and the strength is apparently provided by a combination of flexure and tensile tie action in the link slab reinforcement, aided by the side fascia panels, which are deeper than the link span.

Shortening of the bridge length as a consequence of the fault slip has largely been accommodated by reduction in the distances across the movement joints. Note that the movement joints had not yet been installed when the earthquake struck.

At the abutments, damage is similar to that at internal supports, being largely confined to EDU support blocks, transverse shear restraint blocks and beam ends. However, additional damage was caused to abutment back walls by impact as the bridge was driven towards Istanbul.

The hollow reinforced concrete pier stems were largely undamaged, though a number of piers have small but significant tilts or rotations, particularly where the fault crossed the bridge. Some piers have twisted about the vertical axis by approximately 4 degrees.

With six exceptions, damage to the foundations is minor. Damage to the six exceptions include significant pile cap cracks, and plastic hinging to the piles, as a consequence of gross ground displacements in the vicinity of the fault movement. Where the foundations were rotated by proximity to the fault, damage to the piles was severe.

Displacement demand and near faults effects. Intensive site and theoretical seismological investigations indicated that the redesign input ground motion should be characterized by the following properties, characteristic of a 2000 year return period ground motion (Faccioli et al., 2002):

Design peak ground acceleration (PGA)	0.81g
Design peak spectral acceleration (5% damping)	1.8 – 2.0g
Design peak spectral displacement (5% damping)	600 mm
Consideration of near field directivity effects	

The latter point meant that velocity pulses and fault slip should be considered. A ground permanent deformation of up to 500 mm was estimated to be possible during the design life of the bridge (an average of 5 mm per year over a period of 100 years). Any design action to be considered for the analysis of the viaduct should therefore be based on these assumptions, and in addition should try to satisfy the following conditions, which characterize the Duzce fault and the location of the viaduct:

the magnitude of the earthquake should be of the order of 7 ÷ 7.2;

the earthquake fault rupture should be strike slip;

the recording site should be located with respect to the epicenter in such a way that the angle between the fault and the line connecting epicenter and location is clockwise and small.

With the fault locked until fracture occurs, the question arises as to how will the fault movement be distributed with distance from the fault (which passes through the bridge). Clearly two points some kilometers away from the fault on either side will move relatively by the fault movement, but two points a few meters on either side of the fault will essential experience no relative movement until the fault ruptures. On the other hand, the two points close to the fault will experience the full fault dislocation during the fault rupture as an essentially instantaneous relative displacement, whereas the two points kilometers from the fault will see no additional relative displacement during the fault rupture.

If the relative displacement develops rapidly with distance from the fault, then most of the pier/bearing systems will need to be designed to accommodate the additional displacements due to tectonic movement. However, if the tectonic displacements develop only slowly over a number of kilometers from the fault, then until the fault fractures, the piers of the bridge will maintain their current relative locations as tectonic movements develop. When the fault ruptures, relative displacements will develop only in the immediate vicinity of the fault, affecting bearings in the immediate vicinity.

Both theoretical considerations and measurements of relative displacements of the pier bases supported slow development of tectonic relative displacement with difference from the fault, indicating that the abutments moved closer together by an amount at least equal to the fault dislocation of 1.5 m. This could not have happened if the ground at the abutments had already been displaced by the full relative tectonic displacement. As a consequence, only a small number of piers would need to be designed to accommodate the sum of vibrational and dislocation displacements.

It could also be argued that it is unlikely that full vibrational and dislocation displacements will be additive, as it implies that the dislocation occurs before the vibrational peak (which may be possible, but is uncertain), and that the vibrational response, which is dependent on the development of resonance, is unaffected by the fault dislocation. It would seem probable that the process of fault dislocation would act to damp out vibrational response. As this controversial issue could not be resolved with certainty, the bearings in the vicinity of the fault are to be designed for the full combination of vibration and dislocation.

Intervention strategy. It was immediately apparent that the existing detail for supporting the simple spans on the pier heads would not provide adequate displacement capacity and that continuity over supports would be necessary. Preliminary redesign focused on making the 10-span bridge segments fully continuous and using reduced numbers of large capacity seismic isolation bearings. The process of creating continuity involved casting a new prestressed diaphragm beam at each internal support, of sufficient width to capture the end 600 mm of the beams of the two adjacent spans. Longitudinal continuity between the beams and the new diaphragm beams was achieved partly by shear friction, based on the effective prestressed force, and partly by dowels drilled into the beam end and the side faces. At internal supports, the diaphragm width is 3.6 m, resulting in a significant additional mass to the superstructure of about 10%. Two isolation bearings would be inserted between the diaphragm and the pier head. Placing the diaphragm beam results in very little difference to the way dead load is supported. Live loads are supported by fully continuous action, well within the shear friction capacity of the beam/diaphragm connection. The critical design case for the connection was found to be differential thermal effects resulting from diurnal temperature fluctuations in midsummer. At movement joints, separate prestressed diaphragm beams of reduced width were provided at each segment end.

Normally it can be assumed that the force can be kept to a desired maximum level by seismic isolation, at the penalty of increased displacements. In the case of the Bolu Viaduct, however, the mass of the piers is typically larger than that of the superstructure, as a consequence of the tall piers and the short spans (20.5 MN compared with 14 MN for the tallest piers). With a seismic isolation system placed between the superstructure and the pier head, only the superstructure mass is isolated, and pier moments resulting from pier self-mass are unreduced. As the seismic intensity increases, the pier moments and shears must therefore also increase.

A second concern with seismic isolation was the level of displacement that might develop between deck and pier head. In the normal seismic isolation approach, bearing displacements will be less than the 5% spectral displacement for the isolation period, as a consequence of structure flexibility and additional damping provided by the isolation system. The displacements would be much less than the peak 5% spectral displacements. However, with the high pier mass, the response is essentially a two-mass system, and in the second mode the pier head and the superstructure move out of phase. In this case the bearing displacement can be significantly higher than the spectral displacement for the second mode period. Also, it was clear that the argument related to the first mode, expounded above, was a simplification, since the high mass of the pier would result in the effective height being some distance below the superstructure, and hence the bearing displacement could still exceed the spectral displacement in the first mode alone.

On the contrary, the displacement capacity of the taller piers would exceed the maximum that could possibly develop in the design level intensity. Actually, Moment-curvature analyses were carried out to determine capacities corresponding to different limit states, using material strength data recorded during construction, and without using strength reduction factors. The results of these analyses indicated nominal shear capacities in excess of nominal moment capacities and longitudinal and transversal displacement capacities at the serviceability limit state exceeding 1.0 m and 0.6 m respectively for pier heights greater than 40 m (about 90 % of the cases).

A third concern was related to the specific kind of devices selected by the client, i.e. friction pendulum isolation bearings, for which an appropriate consideration of axial force variation effects may be needed (Calvi et al., 2004). These effects may not be significant for what concerns variation of the displacement demand, but may induce important increment of shear, bending and torsional moment demand on the piers. For the case of the Bolu Viaduct, however, the fundamental parameters that may amplify, or reduce, these effects are rather favorable, as discussed below.

Ratio between deck and pier mass: a significant variation of the shear force transmitted from the deck to the pier may result in strongly attenuated effect at the pier base when the ratio of the pier mass to the deck mass is high.

Aspect ratio of the deck: for the same level of horizontal force, the axial force variation possibly induced by the horizontal acceleration is higher for a deck section relatively larger and for devices relatively closer one to each other. The case of Bolu is again favorable, with a ratio between bearing distance and horizontal forces couple around 5.

Radius of curvature of the viaduct: a curved bridge may result in higher effects, due to the interaction of vertical and horizontal response. The Bolu viaduct is relatively straight.

Intensity of the ground motion: relatively high horizontal peak ground accelerations may induce more significant effects, like in the present case, where a PGA in excess of 0.8 g is assumed.

4.2. Design of the Rion – Antirion cable stayed bridge

The Rion Antirion bridge is located in Greece, between the Peloponese and the continent, over the Gulf of Corinth. The structure will span a stretch of water of some 2500 m with a depth between 60 and 70 m.

No bedrock has been encountered during soil investigations down to depth of 100 m; the sediment depth has been estimated around 500 m; the soil profile is rather heterogeneous, with strata of sand, silty sand, silty clays and clays.

The 2000 years return period design earthquake has been defined by peak ground acceleration equal to 0.48 g and a response spectrum with a maximum amplification plateau at 1.2 g between 0.2 and 1.0 s.

The main part of the bridge consists of three central cable stayed spans of 560 m and two side spans of 286 m, for a total of 2252 m Figure 13.

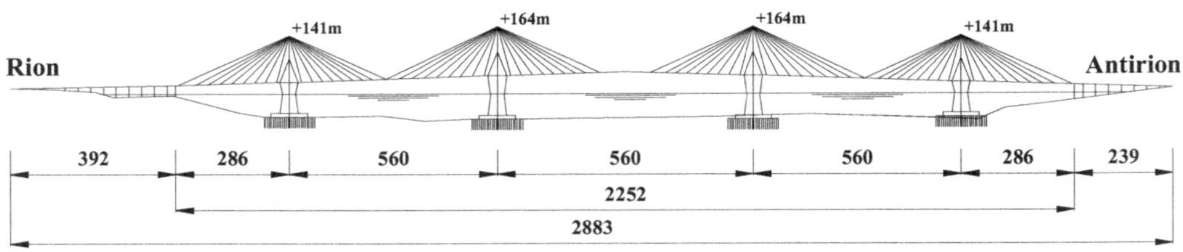

Figure 13 Elevation of the Rion – Antirion bridge

Modal analyses of the structure were performed independently by designers and checkers. In case of a full model, at least 500 modes had to be considered to obtain a total mass participating around 85%. The first modes were typical pendulum modes; the first one had a period of vibration around 7.5 s. The first mode that implied a significant participation of the pier mass was around the 260th, with a period of vibration around 0.6 s. All these essential results were also capture by strongly simplified models.

The foundations are large diameter (90 m) caissons resting on the seabed, where the upper layers have been reinforced with hollow steel pipes, 25 to 30 m long, with a diameter of 2 m, spaced at 7 m distance. Between the top of these inclusions and the base of the foundation, a 3 m thick gravel layer has been inserted, with the purpose of creating a potential sliding surface capable of limiting the shear force transmitted between soil and structure. A fundamental objective of the design was to avoid failure mechanisms involving large rotations, producing extremely large displacements in the high rise pylon, in favor of sliding failure modes, in which case the permanent displacement are of the same order of magnitude along the height of the structure. This is clearly more important than the limitation of shear forces, since the structural response was not significantly different in case of linear or non-linear modelling of the soil.

The assumption of lower, intermediate or upper bound characteristics for the soil response had very significant effects on the structural response. In case of lower bound soil, the displacement demands were normally in the range of three times than those of upper bound soil, with peak values in the range of 4 m between fixed soil and top of the tower. On the contrary, the simulation of traveling effects of the seismic waves had little effects on all design parameters, such as relative displacement demands and variation of forces in the pylon legs.

The cable-stayed deck is fully suspended and behaves like a damped pendulum. Four hydraulic dampers with a capacity of 3500 kN each are connecting each pier head and deck, in addition to a sacrifice steel strut connector, with a 10000 kN capacity, designed resist winds and ordinary horizontal forces, but to break in case of a strong earthquake.

The selection of the characteristic of the damping system was based on extensive numerical analyses, aimed to define the optimal compromise between relative displacement demand between deck and pier and level of shear to be transmitted. In case of assumption of lower bound soil properties, as pointed out by far the most demanding, the force demand with fixed connection was in the order of 70 MN, the displacement demand without connection in the order of 3 m. The use of hydraulic dampers should allow a limit force up to 14 MN, with displacements around 2 m and velocities around 1.5 m/s.

Each pylon is composed of four square legs, with side of 4 m, joined at the pier head to form a monolithic structure where the cables are anchored. The configuration with four legs obviously induce a potential for a vertical stress in one leg larger than four times the gravity stress, since three legs can go into tension. This potential for a high axial stress strongly reduces the ductility and displacement capacity of the leg, where a potential formation of plastic hinge is considered in case of strong earthquake. As a consequence, the gravity stress had to be kept considerably low. The relative displacement demand was estimated around 0.75 m in case of lower bound soil, compatible with the displacement capacity in case of a gravity axial load ratio of approximately 0.1, resulting in a maximum axial load ratio during the design earthquake around 0.5.

Conclusions

Although the consequences still has to be thoroughly explored and exploited, the recent experience and research developments on bridge response, design, assessment and strengthening fully recognize the displacement demand and capacity as the key design parameters. In this picture, the capacity design principles, commonly used also within a force-based approach, will maintain their fundamental role of protecting brittle failure modes in favor of mechanisms able to provide significant displacement capacities and effective energy dissipation. Strength evaluations will be used essentially to compare different damage and failure modes, and therefore to impose or to assess the post elastic mechanism. The emphasis on displacement will require important revision of the representation of the input ground motion, to describe the structure displacement demand as a function of input parameters (magnitude, source mechanism, hypocenter depth, distance between the site and the fault plane). The effects of non-synchronous input ground motions may also find different logic of representation, since they appear to be more closely related to displacement than to acceleration demand.

Various forms of isolation, such as controlled ground layers (in line with the design of the Rion – Antirion bridge), rocking at the foundation level, simple or multiple rocking along the height of the piers, or more conventional deck isolation obtained inserting appropriate devices between pier and deck are likely to become standard practice.

References

Ambraseys, N. and Douglas, J., (2000). *Reappraisal of the effect of vertical ground motions on response*, ESEE Report No. 00-4, Imperial College, London.

Bommer, J. J., Elnashai, A. S. and Weir, A. G., (2000). Compatible Acceleration and Displacement Spectra for Seismic Design Codes, *Proc.12th WCEE*, Auckland.

Calvi, G. M., Ceresa, P., Casarotti C., Bolognini, D. and Auricchio, F., (2004). Effects of axial force variation on the seismic response of bridges isolated with friction pendulum systems, *Journal of Earthquake Engineering*, Vol. 8, no. Special Issue 1, 187-224.

CEN, (1998). EC8, Design Provisions for Earthquake Resistance of Structures, Pub. ENV-1998-2, Comite Europeen de Normalization, Brussels, Belgium.

Elnashai, A. S. and Papazoglou, A. J., (1997). Procedure and spectra for analysis of RC structures subjected to strong vertical earthquake loads, *Journal of Earthquake Engineering*, 1(1), 121-155

Faccioli, E., Paolucci, R. and Rey, J., (2004). Displacement Spectra for Long Periods, *Earthquake Spectra*, Vol. 20, no. 2, 347-376.

Faccioli E., Paolucci, R. and Pessina, V., (2002). Engineering assessment of seismic hazard and long period ground motions at the Bolu viaduct site following the November 1999 earthquake, *Journal of Seismology*, Vol. 6, No. 3, 307–327

fib TG 7.2, (2003). Displacement-based seismic design of reinforced concrete structures, *fib Bulletin 25*, Lausanne

Ghobarah, A. and Elnashai, A. S., (1998). Contribution of vertical ground motion to the damage of RC buildings, *Proceedings of the Eleventh European Conference on Earthquake Engineering*, Balkema, Paris.

Gulkan, P. and Sozen, M. , (1974). Inelastic Response of Reinforced Concrete Structures, *ACI Journal*, Vol. 71, 604-610

ICBO, (1997).Uniform Building Code UBC'97, *International Conference of Building Officials*, Whittier.

Kawashima, K.,(2000). Seismic design and retrofit of bridges, *Keynote address, 12th World Conference on Earthquake Engineering*, Auckland.

Mitchell, J. K., (2000). Kocaeli, Turkey, earthquake of August 17, 1999: reconnaissance report, *Earthquake Spectra*, Supplement to Vol. 16

Newmark, N. M., and Hall, W. J., (1973). Procedures and criteria for earthquake resistant design, building practice for design mitigation, *Building Science Series 45, National Bureau of Standards*, Washington, 209-236

Park R., and Paulay, T., (1993). *Reinforced Concrete Structures*, Jonh Wiley & Sons, 1975

Pecker, A., (2003). Aseismic foundation design process, lessons learned from two major projects: the Vasco de Gama and the Rion Antirion bridges, *ACI International Conference on Seismic Bridge Design and Retrofit*, La Jolla, 2003

Pinto, P.E:, Lupoi, A., Franchin P. and Monti, G., (2003). Seismic design of bridges accounting for spatial variability of ground motion, *ACI International Conference on Seismic Bridge Design and Retrofit*, La Jolla.

Priestley, M. J. N., Calvi G. M., and Kowalsky, M. J., Direct Displacement-Based Seismic Design of Structures, IUSS Press, Pavia, in preparation.

Priestley, M. J. N., (2003). Myths and Fallacies in Earthquake Engineering, Revisited, *The 9th Mallet Milne Lecture*, IUSS Press, Pavia

Priestley, M. J. N. and Calvi, G. M., (2003). Direct displacement-based seismic design of concrete bridges, *ACI International Conference on Seismic Bridge Design and Retrofit*, La Jolla.

Priestley, M. J. N. and Calvi, G. M., (2002). Strategies for repair and seismic upgrading of Bolu Viaduct 1, Turkey, *Journal of Earthquake Engineering*, Vol. 6, no. Special Issue 1, 157-184

Priestley, M. J. N., (2000). Performance Based Seismic Design, *Keynote address, 12th World Conference on Earthquake Engineering*, Auckland.

Priestley, M. J. N., Seible, F. and Calvi, G. M. (1996). *Seismic Design and Retrofit of Bridges*, John Wiley and Sons, New York

Priestley, M. J. N., (1993). Myths and Fallacies in Earthquake Engineering – Conflicts Between Design and Reality, *Bulletin of the NZ National Society for Earthquake Engineering*, Vol.26., No.3, 329–341

Seismic Design and Retrofitting of Reinforced Concrete Bridges, (1994). *Proceedings of the International Workshop*, Queenstown, University of Canterbury, New Zealand.

Seismic Design and Retrofit of Reinforced Concrete Bridges, (1991). *Proceedings of the International Workshop held in Bormio*, Univerità degli Studi di Pavia, Italy.

Sextos, A. G., Pitilakis K. D., and Kappos, A. J., (2003). Inelastic dynamic analysis of RC bridges accounting for spatial variability of ground motion: site effects and soil–structure phenomena, *J. of Earthquake Engineering and Structural Dynamics*, Vol. 32, 607–652

Shibata, A., and Sozen, M. A., (1976). Substitute Structure Method for Seismic Design in Reinforced Concrete, Journal of the Structural Division, ASCE, Vol. 102, No. ST1, 1-18

Somerville, P., (2003). Characterization of near fault ground motions for design, *ACI International Conference on Seismic Bridge Design and Retrofit*, La Jolla.

Teyssandier, J. P., (2003). The Rion Antirion bridge design and construction, *ACI International Conference on Seismic Bridge Design and Retrofit*, La Jolla

SEISMIC TESTING

Pierre Sollogoub[1]

[1] Research Director. CEA/DEN/DANS/DM2S, Commissariat à l'Energie Atomique Saclay, France
pierre.sollogoub@cea.fr

Abstract. This lecture presents the different seismic testing devices, their advantages and limits together with the last developments in this very active field.

1 Introduction

1.1 General

It is well known that seismic engineering is strongly based on experience; first seismic codes were based on horizontal force coefficients determined empirically from post-earthquake observations. Later (in the 30s of past century), when acceleration measurements became available, these coefficients were correlated to accelerations and, then, to response spectra. The behaviour coefficients, which play an important role in design, were calibrated in same way, and structural types with globally good behaviour were assigned a higher coefficient. In similar way, for non structural elements, damage observation oriented the development of design approaches to more vulnerable items, such as thin tanks for which important damages, often with significant safety and economic consequences, can be observed during earthquakes.

The assessment of the seismic vulnerability of structures is a very complex issue due to the non-deterministic characteristics of the seismic action and the resistance of different structural elements, and the need for an accurate prediction of the seismic responses for levels near collapse, well beyond conventional linear behavior. Satisfactory numerical modeling (based on the finite element method) depends largely upon an appropriate calibration of mathematical models able to represent all structural components, including joints. Seismic vulnerability assessment and design of retrofitting solutions for existing structures can be subsequently carried out using calibrated numerical models.

Early, the need of "validation" of analytical approaches or the necessity to investigate new situations pushed the development of testing devices for in-situ or laboratory experiments (Sieberg, 2005); validation of analytical approaches includes numerical values of different parameters, such as damping, global or local rigidities, mode shapes…

In the 60s and 70s, many shaking tables were constructed, mainly in Japan, US and Europe for R&D programs and to perform qualification of equipment for nuclear, defense and transportation engineering. At the same time was developed the pseudo-dynamic testing using reaction walls and allowing large scale testing of buildings and structures. Testing involving soil-mechanics were performed with centrifuge on which special devices were added to apply lateral accelerations simulating earthquake excitations.

During the last decade, it appears that an important testing effort, coupled with analytical and simulation development, is necessary to significantly decrease earthquake induced damages, in terms of death toll and economic losses. Important programs are implemented in USA, Japan and Europe: they include development of testing facilities and improvement of control and capabilities of these facilities.

This chapter will present the different seismic testing devices, their advantages and limits together with the last developments in this very active field.

In this first part, we will present in very general form the different possible seismic test their aim and we will present the similarity laws needed when using scaled model will be then discuss

1.2 Different categories of seismic tests

This part will list the various means which can be considered as experimental, in the sense that they provide results about some loading of a prototype or a model; the results may be qualitative (measurements) or quantative; the loading may be artificial or natural, static or dynamic, comparable or not to a seismic input. The output of the test is to provide qualitative or quantitative data to assess the seismic behaviour of a structure or for calibration of numerical models and , in general, to decrease the uncertainties.

1. ***Real earthquake experience*** feedback. As said before, this technique is used from the beginning of earthquake engineering. It is still intensively used to check and calibrate seismic design and construction practices, giving invaluable information on "what happens" during such event. It is the basis of determination of the Intensity, which is used to correlate damages and vulnerability and losses predictions. It is used for buildings and equipment. In this last case, a systematic gathering of post seismic information in USA is the base of a qualification approach for nuclear equipment (SQUG). The incomparable interest of the technique is that "the real situation" is tested. Among the drawbacks, we can mention the unpredictiveness of the occurrence time, the type of input signal, the amplitude… and the lack of control of the situation. As mention above, it seems clear that the only feedback experience is not sufficient to improve significantly the earthquake engineering. Nevertheless, it is very useful and after each significant earthquake, many post earthquake missions gather invaluable data.

2. ***In-situ tests***. The actual prototype, in its definitive location, is excited either naturally by ambient vibration or by artificial excitation, with centrifuge or hydraulic exciter, outside explosion… The great advantage is to test the real structure in actual conditions. At the opposite, it is not possible to apply strong input energy, especially if the structure will be used after. This approach is intended to validate frequencies and mode shapes of the construction and to give some information on damping. A popular device for excitation is the unbalanced exciter, which apply a sinusoidal force with frequencies up to about 20Hz. Coupling two devices may generate sinusoidal forces in any direction, by action on phase differences. The equipment can be very small and light; with a sinus input, even small forces may capture the resonant frequencies. In Japan, large capacity vibrators, with a force up to 1.5MN are routinely used for "validation" of buildings in nuclear power plants.

3. ***Static test.*** It can be useful either for characterisation of a critical part or for determination of some global behaviour parameters (by simulating a pushover analysis, for instance)
4. ***Shaking table test*** will be discussed in §2
5. ***Pseudo-dynamic test*** will be discussed in §3
6. ***Centrifuge test*** will be discussed in §4

1.3 Objective of test

The objective of seismic test can be the following:

Qualification of equipment under seismic loads, to ensure that it will be "functional" during or after the earthquake. This is often done for equipment in critical facilities, either for safety reasons –nuclear or chemical plants- or for insuring the "life of the city" – telecommunications, hospital… Test is performed following standards, such as IEEE,1987. The instrumentation is usually minimal and the output of the test is Yes or No, the equipment follows the requirement. Usually the test is performed on a shaking table, which reproduces "completely" the seismic excitation. The scale one equipment should be preferably tested. The environment of the equipment must be reproduced, as far as possible; for instance the piping nozzle loads on pumps may have a very important impact on pump operation during earthquake.

Validation of analytical model of the structure or the equipment. This can be the validation of the "whole" model or only of a part of the model. The latter may occur when it is known that some parameter drive the response, such as the boundary conditions, the damping, the value of a given modal frequency. This test is usually performed on the prototype, if possible, or on a model. The prototype must be mounted on site in its actual environment. In that case, the excitation is clearly limited to a low value in order not to damage the facility where the equipment is located. The excitation is either natural, vibration due to the noise – natural or artificial – or through a vibrating device: rotating masses or mass excited by an actuator, blast… It is possible to test only a part of the structure, which has the strongest influence on the seismic response. The test may be dynamic or even static to reproduce the cyclic behaviour of the "critical" part. In parallel to the test, a computational model of the structure under study must be available; the test will validate some of the hypothesis and, reciprocally, the model will participate to the design of the test. This duality between test and computational model is a key question of seismic testing.

Validation of rules for norms and standards. The aim can be to demonstrate the adequacy of some prescriptions of the code, such as reinforcement detailing rules, or more globally, to assess the adequacy of the overall design.

Validation of a "specific design principle", for a "special" construction, in particular when standards are lacking, or special device, such as aseismic pad or damper. The objective of the test (or set of test) is to demonstrate that the device will behave as planned. This is different from the qualification test because in the latter case, the tested item is the total equipment or structure. In the former, it may be only a part of the total structure, but the most sensible. For instance, validation of a rubber based aseismic pad may consider the following tests:

Determination of the constitutive law of the rubber

Ageing properties of constitutive materials

Determination of the behaviour of a single pad, under cyclic loads and up to failure.

Validation of the behaviour of a simple structure or equipment isolated with the device; demonstration of the "correct" behaviour: ultimate state, effect of vertical stiffness, in-structure floor response spectra....

As in the previous case, a computational model of the device must be constructed; the final product of the procedure is a validated computational model.

Research and development tests. The objective is to simulate experimentally a specific subject, as non-linear overall behaviour, to understand the ultimate capacity of structure or equipment, or part of structure... As previously a numerical model of the tested element is at the center of this approach; the test will check the representativity and the accuracy of numerical models and will allow the calibration of model parameters. The main objective of the R/D program will be to validate totally or partly the model. In all the cases a non-linear model is necessary to design correctly the test and to calibrate the level of excitation during the test sequence. An important aspect of tests is related to the confidence brought by a well conducted test.

A specification of the test must be defined; It must contain all the elements necessary for the definition of the test, such as:

Definition of the objective and scope of the test

Definition of the model and similarity law, if relevant

Predictive analyses of the behaviour, quantification of levels for input and outputs

Definition of the instrumentation

Propositions for test sequence

Definition of the data acquisition

Proposals for post-test interpretation

1.4 Similarity laws

Many tests cannot be performed on scale 1 structure; "scaled" models must be designed in order to reproduce the phenomena that are looked at. In this part we will give some basic elements, in a simple way, of the similarity laws usually applied in seismic testing.

Let assume a “real” (scale 1) structure, the prototype, which dimensions are noted L. The reduced scale model has dimensions noted L’. In the following the physical values of the model will be noted with a prime: ‘. The scale parameter, λ, is, by definition such as:

$$L' = \lambda . L$$

Usually, λ is less than one and it is often written as:

$$\lambda = 1/n ,$$

Where n is a real greater than one.

The first similarity condition is called the “**acceleration similarity**”. In this case we want to assume the equality of stresses between the prototype and the model.

The following equations can be written.
Stresses under weight can be writen as:

$$\frac{mg}{S}$$

The mass is proportional to the product of the volumetric mass of the constituent material by the cubic power of a representative length, L and by the acceleration of gravity. This lead to:

$$.L.g \qquad (1)$$

where the sign “=” must be understand as “proportional to”. Writing this equation for prototype and model and equating the stresses lead to:

$$.L.g \qquad '.L'.g'$$

As the model and the prototype are assumed to be “similar”, that is to have the same geometry and “constitution” the above equation is equality. Assuming the same acceleration of gravity, we get:

$$.L \qquad '.L'$$

or

$$' \qquad /$$

This means that either, the material must be changed, or, the mass of the model must be increased, by adding “dummy” masses. The former case is rarely used, because it is difficult to get a corresponding material. In the latter case, the total mass of the model must be:

$$m^{*} \quad m.\;^{2}$$

instead of:

$$m' \quad m.\;^{3}$$

obtained by similarity. The mass of the model must be multiplied by 1/λ, or n.

Under earthquake acceleration (γ) similar equations can be written:

$$' \quad \frac{m'.\;'}{s'} \quad \frac{^{2}m\;\;'}{^{2}.s} \quad \frac{m.\;\;'}{s}$$

If equal accelerations are applied, the stresses under earthquake excitation are identical. In similar way, relations between other quantities, such as velocity, displacement, frequency, force… can be derived. They are reported in Table1. We can notice that the frequency is multiplied by the square root of n and that the time is "divided" by the same quantity: or that the time is condensed, the signal applied is shorter than the real one.

The advantage of this case is that the stresses are correctly reproduced, but an increase of masses is requested. It can be significant and may be the limiting condition for the model design. This is often used for test of civil engineering structures, where the conservation of stresses may be important (the stress due to the weight is important!) and where is it possible to add masses for instance on floors. Another feature is the "contraction" of time and the subsequent increase of frequencies. If the increase of mass is limited, the change of material may be a solution, when the precise investigation of material behaviour is not the main problem.

Then second condition is called "**velocity similarity**", where only seismic stresses are kept and stresses due to weight are no more kept. Under earthquake loads, it can be written, in comparison to the equation (1):

$$.L.$$

for the prototype, and the same equation, with ' for the model.

Equating the stresses gives:

$$.L. \qquad '.L'.\;'$$

Assuming the same material, it leads to:

$$' \quad -$$

From this relation, we can derive relation for time:

$$' \quad \frac{L'}{T'^2} \quad \frac{.L}{T'^2} \quad - \quad \frac{L}{.T^2}$$

From which, we get:

$$T' \quad .T$$

For velocity, we have:

$$V' \quad \frac{L'}{T'} \quad \frac{.L}{.T} \quad \frac{L}{T} \quad V$$

or

$$V = V'$$

This explains the name of the similarity law. Concerning the stresses due to own weight, the relation is:

$$'_{w} \quad \cdot_{w}$$

The advantage of this case is that no additional mass is needed but the stresses due to weight are not reproduced correctly. This case is often used when the stresses due to weight are small compared to those due to seismic loads, such as in a thin tank or more generally for equipment, where it may be difficult to add masses.

The "**centrifuge similarity**"

Some cases may require a perfect stress similarity for weight and seismic loads, without the possibility of adding masses. The typical situation is found in problems involving mechanical properties of soil, such as soil-structure interaction including or not liquefaction. A solution is there to change the acceleration of gravity. The following development can be made:
The stresses due to weight (equation 1) are:

$$.L.g$$

Equating the stresses in prototype and model and using the same material, lead to:

$$g' = g / \lambda$$

There are equal in the model and the mock-up.
The modification of gravity can be obtained by using a centrifuge, which gives an apparent acceleration of gravity equal to:

$$g' = \omega^2/R$$

where ω is the angular velocity and R the radius of the machine. More details will be given in Chapter 4.
The Table 1 summarizes the relationship between the mock-up and the model for the three mentioned similarity laws.

Table 1. Similarity laws

	Acceleration similarity	Velocity similarity	Centrifuge similarity
Displacement	λ	λ	λ
Velocity	$\sqrt{\ }$	1	1
Acceleration	1	$1/\lambda$	$1/\lambda$
Mass	^{2}	^{3}	^{3}
Density	$1/\lambda$	1	1
Weight	^{2}	^{3}	^{3}
Force	^{2}	^{2}	^{2}
Time	$\sqrt{\ }$	λ	λ
Frequency	$1/\sqrt{\ }$	$1/\lambda$	$1/\lambda$
Weight stress	1	λ	1
Seismic stress	1	1	1

These three conditions are to be considered, as a first tentative, when designing a seismic scaled test. The choice of one law will drive the definition of the test and the choice of testing device. It is also possible to adapt the scaling law in order to get a more feasible model and test. But it is important to notice that this do not solve all the situations. If one is interested in determining the damping, should it be internal or "system", a correct determination of similarity would need to know the physical law or equations giving the damping. They do not exist simply, and in many cases, the test is performed to try to find such laws. The scaled simulation of non-linear behaviour of reinforced concrete structures gives another example. It is due to many different mechanisms, such as buckling or slipping of reinforcement, which are very difficult to reproduce correctly on scaled element. Another element to be considered is the behaviour of "scaled" concrete, where all the elements may not be scaled, as the aggregate size or the cement characteristics. Features that are often difficult to reproduce on a model, whatever its scale, are the representativity of "boundary conditions" and the application of loads that operate simultaneously with the earthquake; they must be considered with great care when designing a test. These limitations do not invalidate the scaled test; it is important to have them in mind in order to define the test and to interpret it in optimal way.

Another fundamental point for testing is the instrumentation. It is not the objective of this lecture to give data on this technique, but it is important to have minimum of information about available devices.

The most classical and widely used instruments are:

Accelerometers, measuring the acceleration are either piezoelectric or with force balance. Their size and performances are very wide and, generally speaking, any performance may be obtained, from very small acceleration to very high values, in case of explosive tests, for instance. The usual frequency range is about 0.5-8000Hz for piezoelectric one and 0-500Hz for the other.

Displacement transducers, which measure directly the displacement without double integration of acceleration. Range of measure are 0-100mm to 0-1m for frequencies up to about 70Hz.

Velocity transducers.

Strain gages, which are very common measuring technique. Their sensivity is very high and they can have small dimensions. On the basis of strain gages can be developed different sensors to measure forces, moments, displacements...

In the last years appeared new type of sensors, optical, allowing global measurements of deformation field, by interferometry treatment. This field is in very rapid development and may change some testing practices.

2 Shaking table tests

2.1 General

Use of shaking table is the most natural approach for seismic testing: the prototype is fixed to a rigid plate, moved by hydraulic actuators (sometimes electrodynamic actuators), in order to represent as close as possible the required seismic movement. The model is instrumented and the resulting measures are recorded and processed. The main feature is that a base movement excites the model, exactly as it would be during an earthquake, if the boundary conditions are well

represented. The "counterpart" is the need of important "hardware" and the necessity to have a scaled model, due to the limited mass capacity of tables.
Figure 1 shows the general view of the Azalée shaking table in CEA Saclay

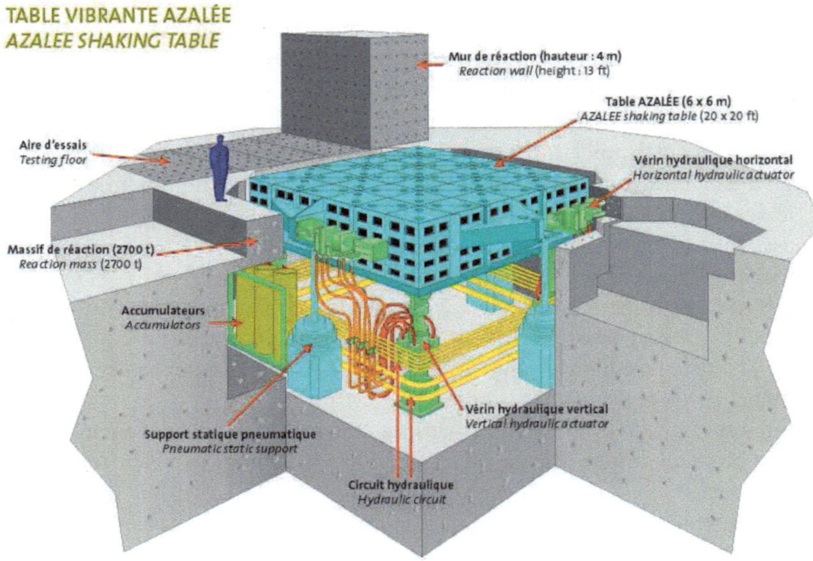

Figure 1. Azalée shaking table

The different parts of a seismic shaking table test facility are:

a rigid table
actuators
a concrete "reaction" mass
an electronic command (hardware part) to drive the actuator,
a computer to generate the program signal

The different parts will be examined in more details.

1. The table: this is a rigid plate made of steel, in the majority of cases, aluminum, in order to save mass and in the same time increase the allowable model weight (Azalée) and concrete (UC Berkeley). For metal case, it is a complex welded structure, which must be rigid and light. Typical frequencies of the empty table are above 70-80Hz. The table is supported by

actuators in a 6dof table, but in other cases they are either rods or hydrostatic oil pads: they must have a high rigidity and a large capacity of displacement with limited distortions.

2. Two types of shakers or actuators can be used:
   electrodynamic shaker;
   hydraulic actuator.

Each actuator has its own advantages and disadvantages:

For electrodynamic shakers, the main advantages are: large frequency range of excitation and especially at very high frequency (up to 2000 Hz) and low distortion. The main disadvantages are: very low stroke (1 or 2 inches), impossibility to operate at very low frequency (under 2 or 5 Hz), their important volume, their cost and the relatively limited load.

For hydraulic actuators (Figure 2), the main advantages are: long stroke (200 mm and over), important force or power and low volume. Among the disadvantages, it can be mentioned: frequency range limited to 100 Hz or 300 Hz and the important distortion. For seismic tests, all laboratories use hydraulic actuators because the low frequency range is important when structural damage is concerned.

Figure 2. General view of a hydraulic actuator (MTS)

A shaking table is moved with one or more hydraulic actuators; their configuration depends on the number of activated degrees of freedom, from one to six. In the latter case, usually 8 actuators are needed. They are powered by oil under high pressure; for that, hydraulic pumps are needed.

The main characteristics of the pumps are the flow rate they can deliver in permanent conditions. For certain transient conditions, accumulators may be needed to supply supplemental flow rate.

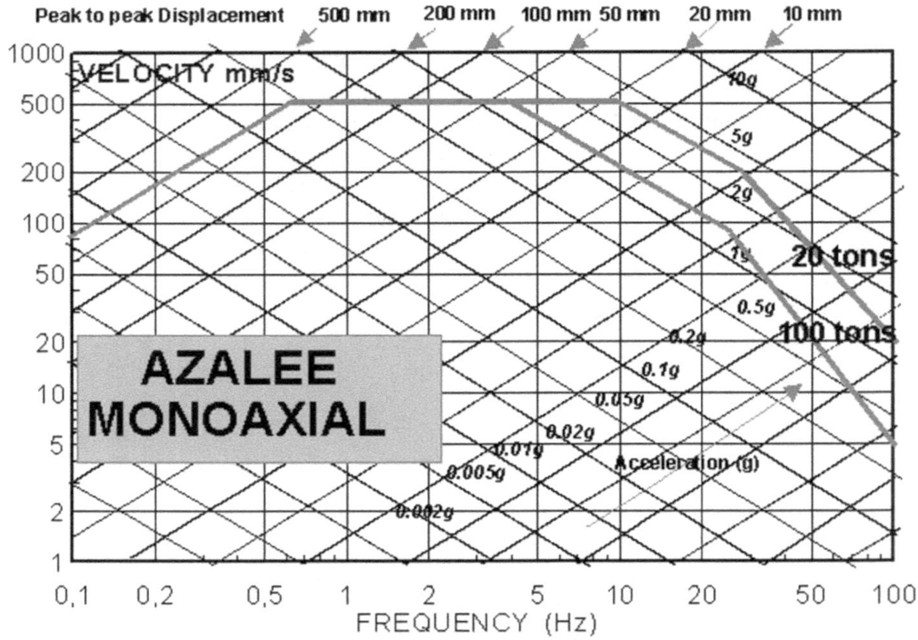

Figure 3. Maximum performance of Azalée table for sine sweep monoaxial excitation

An actuator is composed of a cylinder with bearings, a piston, one or several servovalves and displacement transducers (LVDT). The servovalve is the heart of the system. It acts as a hydraulic amplifier, which supply oil to the actuator as function of an electrical signal. The pressure and the diameter of the actuator control the maximum applicable force and, consequently, the maximum acceleration applicable to a given mass; the flow rate supplied by the pumps or acceptable by servovalve limits the velocity; the displacement is limited by the maximum actuator stroke. The velocity limits imposed by servovalves imply frequency limits: for small actuators, those limits are about 100-150Hz and for large one, 50-100Hz. Another limit is due to the flexibility due to the compressibility of the oil column. Even if oil's modulus of compressibility is high, it induces frequencies about 7-20Hz, depending on the supported mass, for large tables. This frequency is physically "unavoidable" and the control of the system must deal with this phenomenon. The Figure 3 presents the maximum performance of the CEA Saclay Azalée table under sine sweep monoaxial excitation. The three limiting conditions can be clearly seen: at low frequency, displacement limit, at high frequency, acceleration and at median frequency, velocity.

3. Other

The shaking table is supported by a concrete mass, working as a reaction mass. Two possibilities are present: either a mass directly located on soil or a suspended one on springs or similar devices; the frequency of the suspended mass must be low, around 1 to 2 Hz. The former case is the simplest but it may require an important mass; the vibrations transmitted through the soil to the surrounding buildings may be significant. The latter solution decreases the transmitted vibrations, but is more expensive; it is usually considered that the suspended mass should be 20 times the payload mass of the table. In the case of a mass directly located on ground this ration may be multiplied by about 10!

The Figure 4 illustrates the configuration of the 30t capacity table of CESI-ISMES in Italy. The reaction mass is suspended on air-springs.

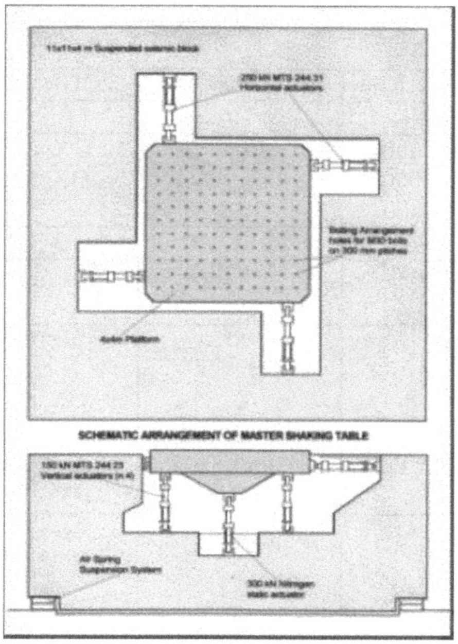

Figure 4. CESI-ISMES table – general configuration, reaction mass (CESI-ISMES document)

Another important part is the **hydraulic system** with pumps, tanks, accumulators and piping systems. These parts are rather conventional industrial components and will not be described more precisely; only some specific points will be mentioned. The pressure of the oil is usually above 200bars (usually 210 and 380bars). As mentioned, the important parameter is the maximum available flow rate: many facilities have accumulators that increase the flow rate during the strong phases of seismic signal; then the duration of the test is limited. As an example, for the Azalée table in CEA, the accumulators can multiply the flow rate by about 4, from 2200liters/mn to more than 8000liters/mn. In the Miki table, all the pipes are not welded, but assembled by specially designed devices. This part needs a significant maintenance effort, which should not be neglected.

Lastly, the data acquisition and processing system is an essential part, mainly computerized. The interface between instrumentation and acquisition, where the measure is conditioned in order to be treated by the acquisition material is to be mentioned.

In the exploitation of a shaking table facility, the "auxiliary" features, such as area for model construction and handling, are an important point to be considered.

For illustration, the Table 2 presents the main characteristics of the 4 shaking tables in the CEA Saclay Seismic Laboratory.

Table 2. Characteristics of shaking tables of the CEA Saclay Seismic Laboratory

Test set up	AZALEE	VESUVE	TOURNESOL	MIMOSA
Number of dof	6	1	3	&
actuators axis and load (kN)	OX = 2000 OY = 2000 OZ = 4000	OX = 350	OX = 100 OZ = 200	OX = 500
Dimensions (m)	6 X 6	3.1 X 3.1	2 X 2	2 X 2
Mass of the table (tons)	25	4.2	1.2	1
Maximum payload (tons)	100	20	10	10
Axis	Triaxial OX, OY, OZ	monoaxial horizontal OX	biaxial horiz.+ vertical (OX+OZ)	monoaxial horizontal OX
Maximum Displacement (mm)	OX and OY 125 OZ 100	100	OX = 125 OZ = 100	12.5
Maximum velocity (m/s)	OX= 1 OY = 1 OZ = 1	1	biaxial OX = 1.4 OZ = 0.7	0.6
Maximum acceleration (g) (max payload)	OX = 1 OY = 1 OZ = 2	1.2	OX = 1 OZ = 1.5	4
Frequency range (Hz)	0-100	0-100	0-100	0-300
Maximum height of the model (m)	12	12	12	12

Some comments on the characteristics can be made: Azalée, the largest table in Europe, has a maximum payload of 100 tons. The maximum stroke of the actuators is +/-0.125m, which was the typical feasible value in the end of 80's. The maximum velocity is 1m/s. The frequency range is typically 0-100Hz, which is representative of seismic movement. Mimosa is a "high frequency" table, with its 300Hz maximum controllable frequency; the maximum stroke is only +/-12.5mm.

4. Control system

An essential element of shaking table test is the control of the table, or the capability of applying the desired input (up to 6 dof) even with a large model. The input is usually defined as imposed acceleration in translation and, possibly, rotation. The controlling parameter is servovalve position. This is basically a control problem which can be solved in different ways. If we consider a simple monoaxial table, the command of the actuator is provided through movement of the servovalve. As the system is more complete and complicated than a single actuator, loop control is necessary: the movement of the actuator is measured by an LVDT and this signal is compared to the imposed signal. The difference (or error signal) is send to the servovalve as shown on the Figure 5.

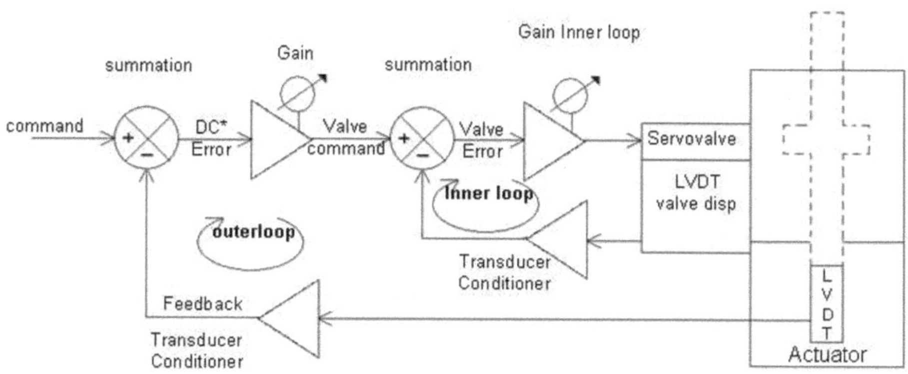

Figure 5. Basic closed-loop control

Inside loops may exist (ΔP feedback) in order to stabilize the servovalve movement due to the oil column frequency. One important thing is that a broad band (typically from .05Hz to 100Hz) control is needed for seismic tests. One given signal has a very broad range of frequency. Therefore, some systems combines information from three signals: displacement, effective at low frequency, velocity at medium frequency and acceleration at high frequency. This control allows a rather "flat" response in a larger frequency range. The control system should also manage the other degrees of freedom (if more than one) and the existing geometric and mechanical coupling between them.

The model, by its mass and rigidity, modifies the system characteristics; in order to cope with this, a preliminary test is performed to determine the new transfer function between table movement and required movement. This is done by applying a white noise excitation and measuring the table response. By Fourier transform, transfer functions can be determined. These transfer function are used to determine an input signal such that the table has the prescribed movement. The test is performed at very low level, in order to avoid damage to the test model, especially if it is a reinforced concrete or masonry one. With the increase in calculation speed, real time control may be envisaged. (see §4).

2.2 Shaking tables in the world

As mentioned above, first shaking tables appeared in the 30s (see for instance Sieberg, 2005), with a very limited capacity (some kg). Larger tables appeared in the 60s in USA (Berkeley), Japan, France (in 1968 in Saclay), Italy (with a table moved on railway in ISMES)...
Many laboratories are equipped with shaking tables. The Table 3 is a list (non exhaustive) of the most known tables. It is derived from OECD/NEA (2004).

Table 3. Shaking tables

Institution	Country	Number of dof	Payload (tons)	Area (m^2)
CEA	France	6	100	36
Hydroproject research Institute	Russia	3	45	36
LNEC	Portugal	3	36	31
Univ. SS Cyril and Methodi	Rep of Macedonia	3	36	25
KFA Juelich	Germany	3	23	25
ENEL HYDRO /ISMES	Italy	6	15	16
Univ BRISTOL	UK	6	14	9

ENEA	Italy	6	9	16
NTUA Athènes	Greece	6	9	16
Ansaldo	Italy	3	6	12
Nishimatsu Contruct. Corp	Japan	6	NA	30
National Inst for Earth. and Disaster Prev.(MIKI)	Japan	3	1200	300
Nuclear Power Eng. Corp (Tadotsu)	Japan	2	1000	225
Public Works Research Inst.	Japan	6	272	64
Aichi Inst. of Technology	Japan	1	136	66
Sanryo Heavy Industries corp	Japan	3	91	36
Hazama cop	Japan	3	91	36
Kumagai-Gumi Corp	Japan	6	64	25
Kajima Corp	Japan	6	46	25
Nat. Research Inst. of Agric. Eng	Japan	3	45	24
Obayashi-Gumi Corp	Japan	3	45	25
Inst. of Machinery and Metals	Korea	6	27	16
Nation Center for research in EE	Taiwan	6	27	25
Fujita Corp.	Japan	1	25	16
NYK Corp.	Japan	6	20	7
Shimizu Corp	Japan	3	20	16
Tobishima Corp	Japan	3	20	16
Taisei Corp	Japan	2	20	16
Hitachi Engineering Cop	Japan	1	20	16
Building Research Inst.	Japan	3	18	12
Kyoto Univ.	Japan	6	14	15
Tonji Univ.	China	2	14	16
NPIC (Nuclear power Institute of China) Chengdu	China	6	60	36
CHRI (Chongqing)*	China	6	40	18
Univ. of Buffalo*	USA	5	50	13
Univ. of Berkeley	USA	6	45	37
US army Civil Research Lab	USA	3	45	13
Univ. of Nevada RENO	USA	2	45	19
Univ. of California San Diego	USA	1	33	15
Wyle Laboratories	USA	2	27	37
Univ. of Illinois URBANA	USA	1	5	14
Univ of Pavia (Eucentre)	Italy	1	60	39

* Facility with two “independent” moveable tables.

The Table 3 gives an idea of the development of this testing technique, indicating only few basic data. Largest tables are, not surprisingly, located in Japan. The Tadotsu table, with 1000t pay load and open in 1984 was the largest table during more than 20 years. It was used mainly for "proving" and R/D tests for Nuclear industry. These last years, a second one directional shaking table has been mounted above the table, in order to obtain very high accelerations for "fragility" tests. The table is now decommissioned.

Another Japanese table is remarquable: it is located in Miki, near Kobe and its payload is 1200t with a 20mx15m (6dof) table. It has been designed and is operated by NIED (National Research Institute for Earth Science and Disaster Prevention. It is important to associate the very advanced performances of the actuators, in the horizontal direction: maximum actuators stroke: +/-1m, maximum velocity: 2m/s. These values are now typically foreseen for new facilities. The Miki table was officially inaugurated in January 2006, and it will allow scale 1 tests for many structures. The Figure 6 is a sketch of the table: many outstanding features are included in the design and construction of the facility. The total oil volume is 750000liters. The mass of the table plate is 700tons. The motivation to start this project was the important damages during the Kobe earthquake. The importance of studying full scale structures subjected to strong movements was then realized.

Figure 6. Miki Shaking table (NIED- E Defense)

More detailed information can be found in CASCADE n°6 (2005) report.

It is worth noticing the very important effort made in USA in order to improve and develop seismic testing facilities under the NEES (Network for Earthquake Engineering Simulation). In the framework of this network an important funding was dedicated to the improvement of existing facilities and to the construction of new shaking tables (as in Buffalo), reaction walls, wave basins, multiaxial loading systems... (see the NEES web site for more details).

2.3 Examples

Some examples will be shown in this paragraph, with the objective of the diversity in tested model and in objectives of test campaign.

Figure 7. Six-story RC frame model prepared for test on Miki Table

Figure 7 shows a 6-story reinforced concrete frame built for test on Miki table, illustrating the possibility of this huge machine. It is recalled that the objective is to fail a scale one structure in order to get data in this field. An important activity on Miki table is foreseen in behaviour of foundations and effect of liquefaction for which a shear stack of very large dimensions is placed on the table.

Figure 8 shows a qualification test for an isolator in an electrical substation; the test was performed on the CESI-ISMES table in Bergamo (Italy). The objective is to demonstrate the acceptable behaviour under earthquake loading. The test follows requirements such as IEEE standards for electrical equipment (see for instance IEEE – 1997) or more generally the IEEE –

1987 for seismic testing of electrical or mechanical equipment. CASCADE2 (2005) and CASCADE3 (2005) give complementary information on this matter.

Figure 8. Qualification test of an isolator (CESI-ISMES- Italy)

The Figure 9 shows a test of a piping system performed on the Azalée table in CEA Saclay. The objective of this R&D test is to quantify the ultimate behaviour of piping systems, widely used in industrial and other facilities and to support changes in Codes and Standards. The conventional seismic design of piping requires to install many supports to "counter balance" seismic loads. But test show that it is almost impossible to fail a system by the "inertial" loads. Moreover, the addition of supports stiffen the line and increases the stresses due to thermal loads, creating a more "critical" situation (see Touboul and al, 2006).

Figure 9. Piping system test (CEA Saclay - Azalée)

3 Pseudo-dynamic tests

3.1 General

A significant limitation to shaking table tests is the limitation of the size of models, even with the largest tables. The desire to test large model pushed the development of PSD (PSeudo-Dynamic test), together with the improvement of numerical and computer techniques. The PSD is an hybrid technique, coupling test and computer analysis. The behaviour of the structure is described by the classical equation of motion:

$$M\ddot{x} + C\dot{x} + R = -M\ddot{X}_e \qquad (3.1)$$

where it is assumed that the structural behaviour has been discretized in suitable manner, such as with a finite element model. In the equation (3.1), x represent the displacement, M is the mass matrix, C, the damping matrix, X_e, the applied base movement and R is the restoring "structural" force. The most difficult part to simulate is precisely this restoring force. Then the PSD will combine test, representing the restoring force, and analysis, representing the other terms of the equation (3.1): the equation is solved numerically, at a given step, using information (forces) from the actuators and giving the displacements for the next step. A common implementation of the technique is schematically represented on Figure 10, with a reaction wall and strong floor, actuators applying forces or displacements to the structure at points where are concentrated masses and a measure and control system.

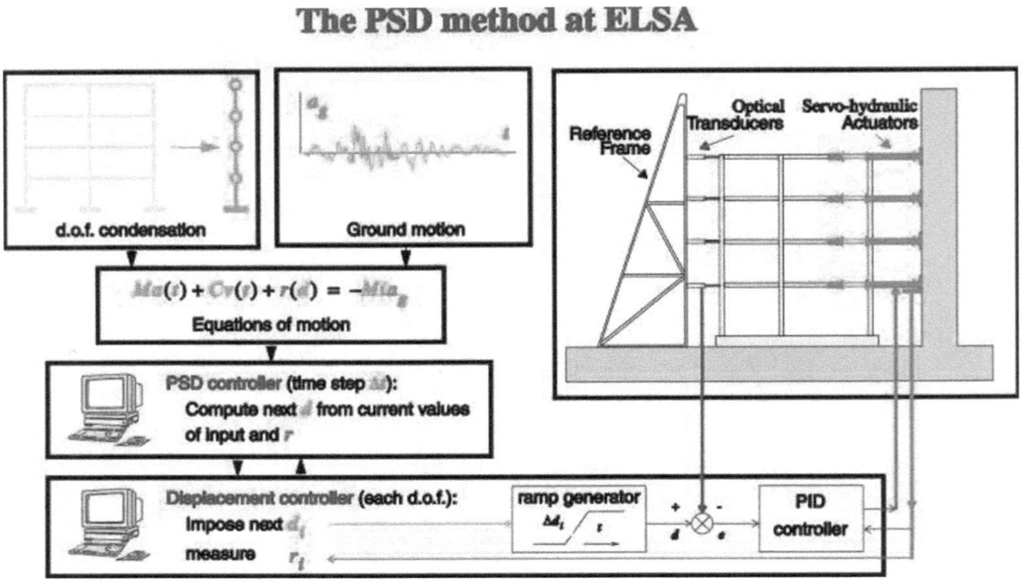

Figure 10. General components of PSD test (from JRC – Ispra, ELSA Laboratory)

3.2 Implementation

The technique was first implemented in Japan in the 70s (Takanashi K. et al. 1975). It was developed in static laboratories where strong walls often exist. Some developments of the technique were implemented in United States, at the University of California in Berkeley and the algorithms and overall control significantly improved. The different systems differ mainly by the implementation of algorithms, measure and control.

The simplest way to understand the numerical part is to consider the classical central difference method to discretize in the time domain the equation (3.1). The development is taken from CASCADE7, 2005. A time step Δt is chosen, and the velocity and acceleration are expressed as:

$$\dot{\mathbf{u}}_t = \frac{\mathbf{u}_{t+1} - \mathbf{u}_{t-1}}{2\Delta t}, \qquad \ddot{\mathbf{u}}_t = \frac{\mathbf{u}_{t+1} - 2\mathbf{u}_t + \mathbf{u}_{t-1}}{\Delta t^2} \tag{3.2}$$

From these equations, it follows:

$$\mathbf{u}_{t+1} = \mathbf{u}_t + \Delta t\dot{\mathbf{u}}_t + \frac{\Delta t^2}{2}\ddot{\mathbf{u}}_t$$
$$\dot{\mathbf{u}}_{t+1} = \dot{\mathbf{u}}_t + \frac{\Delta t}{2}(\ddot{\mathbf{u}}_t + \ddot{\mathbf{u}}_{t+1}) \tag{3.3}$$

The PSD sequence can be written in the following way:

At the end of step i, the structure is at equilibrium
At the beginning if step i+1, the second hand of (3.1) is known (it is the applied input signal)
From the first equation (3.3), the displacement at i+1 is calculated
This displacement is applied to actuators
The restoring forces R_{i+1} are measured
From (3.1) written at step i+1and the second equation (3.3), velocity and acceleration at step i+1 are calculated
Proceed to the next time step

The method "merge" analytical and test results, which may cause systematic errors, which, cumulating with time, may lead to completely wrong results and premature failure of the model. As illustrated by Pegon (2004), this is, somehow, equivalent to solve numerically differential equations with a computer having a limited number of digits.

The fact that u_{i+1} can be obtained directly in the previous development, is due to the explicit character of the central difference scheme. This is very convenient for calculation and for the explanation of the PSD, but this method is not unconditionally stable. For structures having "rigid" parts (which is the case of almost every "real" structure) this requires a small time step and, consequently, the risk of obtaining at some time steps very small displacement, less than the precision of the measurement system. In the ISPRA – ELSA facility, displacement measurements are made by optical device with resolution of about 2μm and a total precision of the displacement control of about 50μm. The displacement is measured relatively to a special rigid structure (see Figure 10), and not relatively to the reaction wall, which may have some flexibility. For some researchers, it is better to use the previous computed displacement than the measured one.

Implicit algorithms are generally superior to explicit in term of stability and energy dissipation at high frequency, but they require iterations, which may apply strong forces (overshooting) to the tested structure. These iterations often require the knowledge of the tangent stiffness matrix, which is not easy to determine by test. Some implicit algorithms were developed using the initial stiffness matrix and leading to unconditionally stable analysis (see Shing et al, 1991), which must be performed with great care, in order not to damage the structure: "overshooting" must be avoided. Combescure et al. (1997) proposed an explicit Newmark type algorithm with splitting of operators, where the linear part is implicit, but without iteration thanks to linearity, and explicit for the non linear part.

Some other features, advantages and drawbacks of the PSD are summarized hereafter:

One possible advantage of the method is the possibility of having time between two time steps to examine in detail the structure (in-depth mapping of cracks…), to adapt the subsequent test sequence (until failure) and, if necessary, add or change instrumentation

On the contrary, the "wait" period, between two time steps: for some materials, relaxation may occur and a long duration of the wait period may change the measured forces and/or displacement. This wait time can be reduced by using the continuous approach (see below). In the conventional PSD approach, each time step includes a hold

period during which measurements (actuators force…) are done and then calculation of the next step displacement; then the displacement is reach with a ramp (see Figure 10).
The main advantage is the possibility of testing scale one models. There is no need to characterize preliminary the structure, the test will do this.
Tests are performed at "reduced" velocity: some viscoelastic materials, with a law depending on velocity, are not well represented. In general, viscous damping, represented by the matrix C in (3.1) is assumed, a priori, and maybe checked a posteriori.
The masses are assumed to be concentrated and in a limited number: this is a good assumptions for many structures, but not for all of them. Non linearities such as gaps are difficult to represent.
The vertical excitation is usually not considered, which may cause some troubles (see Camus experiment). In theory, it is possible to represent it with vertical actuators.
For the conventional PSD method, only static actuators are needed, which simplifies the hydraulic part. As an exemple, the ELSA laboratory in JRC-Ispra has a maximum flow rate of 1500liters/min compared to Azalée which have a maximum permanent flow rate of 2200liters/min and a peak value of 8800liters/min with accumulators.

3.3 Reaction walls in the world

Table 2, taken from OECD/NEA presents some basic characteristics of largest reaction walls.

Table 2. Largest reaction walls in the world.

Institution	Country	Height (m)	Surface (m^2)
EC Joint Reseach Center	Italie	16	281
Building Research Inst *	Japon	25	NA
Hazama Corp	Japon	18	423
Sumitomo Corp	Japon	12	NA
Nihon Univ.	Japon	12	285
Kumagai	Japon	10	NA
Shimizu Corp	Japon	10	NA
Southwrest research Inst	USA	NA	930
Univ. Of Texas Austin	USA	NA	670
Construction Technology Labs	USA	NA	630
Purdue Univ	USA	NA	465
Univ. Of Buffalo	USA	9	453
Univ. Kansas	USA	NA	372
Univ Nebraska	USA	NA	319
Lehigh Univ. *	USA	15.2	381
Univ. Of California San Diego	USA	15	946
Nation. Inst. of Standards and Tech.	USA	14	345

Univ. Of California San Diego	USA	13.3	590
Cornell Univ.	USA	12	300
Univ. Of Minnesota *	USA	12	297
Univ. Nacion. Autonoma de Mexico	Mexico	10	NA
Univ of Nevada RENO	USA	9.5	765

* L-shaped wall

The Table indicate only some "rough" hardware figures; the "quality" of the laboratory relies not only on the hardware, but in the procedures and the implementation of algorithms for the control. Figure 11 presents in more details the characteristics of the ELSA reaction wall, in JRC Ispra Laboratory.

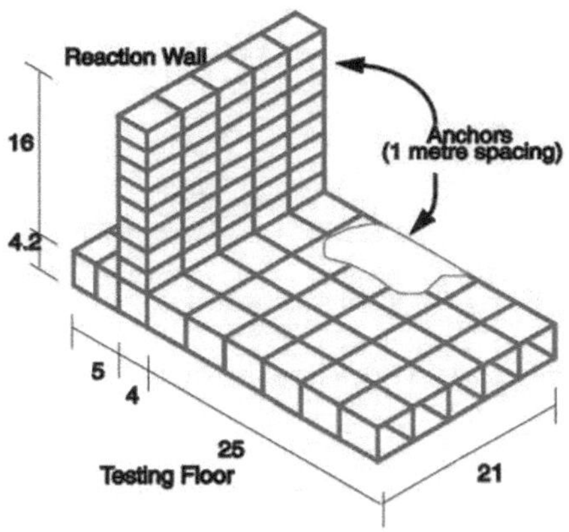

LOAD CAPACITY	REACTION WALL	Bending Moment	200 MNm
		Base Shear	20 MN
	REACTION FLOOR	Bending Moment	240 MNm
	ANCHOR LOAD	Axial Force	500 kN
HYDRAULIC CHARACTERISTICS	FLOW		1500 l/min.
	PRESSURE		210 bar
	ACTUATORS	Load (MN) Stroke (m)	(0.5 - 1.0)) (0.25) - (1.0)

Figure 11. PSD method implementation (from JRC – Ispra, ELSA Laboratory)

The figure highlights some important points, as the necessity of a strong floor and the positioning of anchors, essential for tests. Some reaction walls are L-shaped, allowing easy tests in two horizontal directions. The facility in the University of Berkeley have a "reconfigurable" wall with reinforced concrete blocks which can be fixed together by post tensioning, with different geometries.

3.4 Examples

Some illustrations of PSD application are given below.

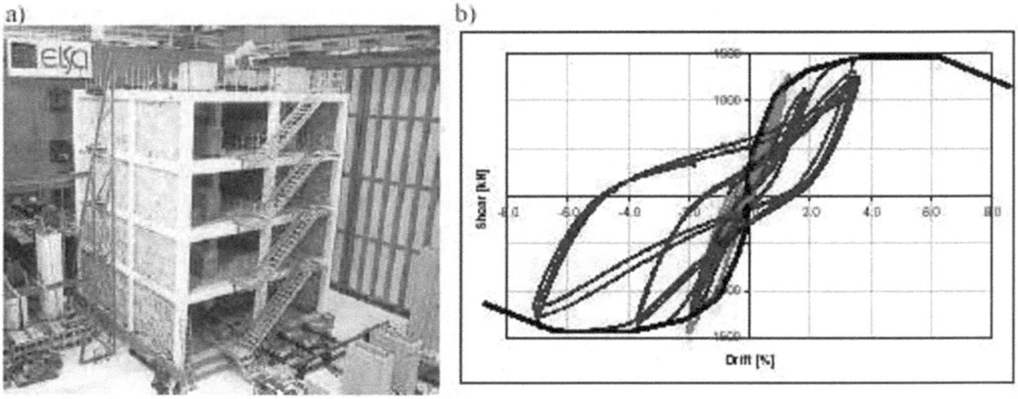

Figure 12. R/C structure tested ELSA laboratory (from JRC – Ispra, ELSA Laboratory)

The Figure 12 shows a test performed in JRC-Ispra on a 4-storey R/C structure designed according to the Eurocodes 2 and 8. Tests were performed at different levels; at the level corresponding to the design earthquake, some low damage was observed: cracks in critical sections of beams and at the base of first story columns. Before the "collapse" test, intermediate tests were performed with masonry infills in different configurations. The right part of the Figure 12 shows the first story displacement versus load for the ultimate test (Negro, 1996). The excitation was applied in one direction in a triangular shape for the last test.

Figure 13 presents a test performed on a torsionally unbalanced building. Actuators (four per storey) act in two directions as shown on the figure. Exterior actuator is fixed to a specific steel structure. Each story has 3 degrees of freedom (two translations and one rotation around vertical axis), which make the control complex. After initial tests, retrofit with carbon fiber has been tested.

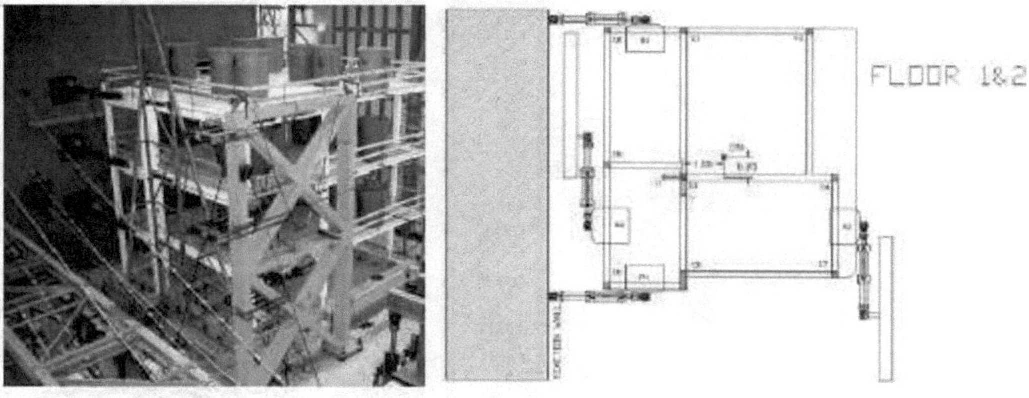

Figure 13. 3D tests on a torsionally unbalanced structure (from JRC – Ispra, ELSA Laboratory)

4 Centrifuge tests

4.1 General

Geotechnical centrifuges are used for research in geotechnical science; they generate an artificial gravitational field in order to reproduce the stress state in the soil, according to the "centrifuge similarity" presented in §1. Centrifuges are used to study the effect of gravity on soil samples or small-scale models of structures. First idea of application of this technique was proposed in 1869 by a French engineer, Phillips, and it was put into practice in the 1930s in USA and Russia.

The principle of the beam-type machine, the most common, can be seen on Figure 14. The different parts are a symmetric arm rotating around a vertical axis, with a "swinging" basket at one end in which is located the tested specimen. At the other end, there is a balancing device. The size of specimen is typically around 1m for each direction. The main characteristics of machine are:

i) the nominal radius, distance between the center of payload and the centrifuge axis

ii) dimensions of the model

iii) performance of the machine, which is quantified by the maximum achievable acceleration as function of payload mass and is usually expressed as the mass multiplied by the acceleration, expressed usually in g.

Conventional devices are installed in the "basket" in order to apply static loads to the model; one point is the limited available room. Seismic (dynamic) loads are applied by different means. In the most recent facilities a small shaking table is installed in the basket with hydraulic or electrodynamic actuators. The system is associated with some complex mechanical parts in order to minimize the total unbalance of the basket, to which the centrifuge is very sensitive. Older machines used suddenly released springs, explosives (Luong, 1992) or drop-ball arrangement (Semblat et al, 1998). In the LCPC machine, the seismic signal is actuated by accumulators located in the basket.

Figure 14. General view of a beam-type centrifuge (LCPC, Nantes France)

The control of the centrifuge is obtained by controlling the velocity; the "in-flight" shaking table control is a conventional one. The experiment, including soil is usually placed in a "shear stack" or "laminar box"; an example is presented in Figure 15.

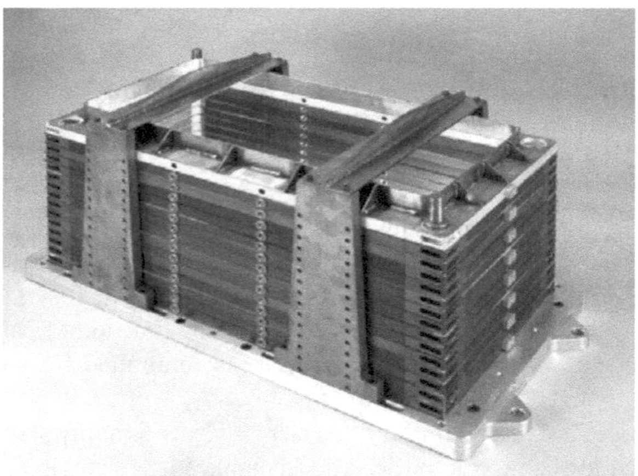

Figure 15. Laminar box for centrifuge (LCPC, Nantes France)

An important aspect of this technique is related to the limited size of specimen, which complicates the experiment. It must be recalled that the geometry, including grain size, is scaled to about 1/100. Usually fine grain sand, or clay, is used for soil which is placed by special "pluviation" system to obtain a very homogeneous medium. Coming back to similarity laws, some behaviours such as dilatancy, strain rate effects, permeability, viscosity of the water are not covered by dimensional analysis and requires a large experience and skill from the laboratory.

4.2 Centrifuge in the world

The Table 3 presents a partial list of centrifuges, giving an idea of existing models; data from CASCADE6, 2005, has been used. Some of them are very recent, or under construction in Japan and in USA, under the NEES initiative. An important point is the capacity of performing seismic tests.

Table 3. Partial list of centrifuges

Owner	Country	Radius (m)	Maximum acceleration (g)	Maximum payload (tons)	Acceleration rating (g.t)	Year of First operation
Delft Geo	NL	6	400	5.5	2200	1989
UCambridge	UK	4.13	150	1	150	1973
UDundee	UK	3.2	150	1.5	150	1999
UManchester	UK	3.2	130	2	260	1971
ETH	CH	1.1	440	2	880	2000
RU- Bochum	D	4.13	250	2	500	1987
CEA-CESTA	F	10	100	1	100	1956
LCPC	F	5.5	200	2	200	1985
PWRI	Japan	6.6	150	5	400	1997
MinTransp	Japan	3.8	113	2.7	312	1980
Kajima	Japan	2.63	200	1	100	1990
Obayashi	Japan	6.6	50	3	700	
NC Univ	Taiwan	3	200	1	100	1995
RPI	USA	3	200	1	100	1989
UCalif	USA	9.14	300	3.6	1080	1988
UColorado	USA	6	200	2	400	1988
USArmy	USA	6.5	350	8	1256	

4.3 Examples

Seismic tests performed on centrifuge are obviously related to soil and geotechnical matters: soil structure interaction, load-deformation behaviour of superficial foundation, study of deep foundations (piles), soil pressure on constructions, soil and structure permanent deformation, liquefaction, interaction of structures with liquefied soil…

Two examples illustrate typical applications of this technique. The first one presented in Gajan et al., 2005 is a test campaign aiming to the determination of load displacement behaviour of rocking shallow foundations. It was tested at the University of California in Davis on a 20g centrifuge accelerator. Different configurations of foundations are placed on a sand layer. The dimensions of the container are 1.75x0.90x0.53m; six to ten models are included in each container configuration. Different loading devices apply shear or normal forces or overturning moments. Two different soils are considered: Nevada sand or San Francisco Bay mud; sand was air pluviated. Quasi static, cyclic loads are applied with small actuators in vertical and lateral directions, coupled or not; displacements at different locations and forces on the actuators are measured. They are processed in order to determine force-displacement curves. Local deformation of soil and footing contact are examined. The effect of embedment was quantified. Many valuable results on the non-linear moment rotation curve were obtained, The main interest of the centrifuge test was to represent the effect of the confining stress, which cannot be represented on a 1g test. Comparable test was performed in France to check the Rion-Antirion bridge (Greece) foundations design.

The second example presented in Lee, 2005, represents the model of a quay wall which was displaced during the Chi-Chi earthquake in 1999, in Taïwan. It was tested on the RPI Centrifuge in USA (Rensselaer Polytechnic Institute) at 120g. The container length is 62cm, representing about 74m of the actual structure (scale 1/120, as shown in §1.4). Water has an important effect during the earthquake and the rate of pore fluid dissipation quantification is important. For this objective, water was replaced by a special fluid with high viscosity in order to catch the correct effects. Displacement and pore pressure were measured. Global deformation of the wall similar to the observed one, has been obtained on the model; a more in-depth understanding of the involved phenomena was achieved.

5 New developments

The field or seismic testing is very active, due to the importance of R&D needed in this field, with the objective of decreasing significantly losses during earthquake events. Many developments are underway, with the objective of improve the testing capacity, with the construction of new facilities and with the development of new techniques. Some of them are listed here.

Real time control of shaking tables; As mentioned, the precise control of tables is a difficult task, when significant non linearities of the structures appear or when precise differential displacement are required. Techniques to improve the control are proposed (CASCADE7, 2005); they modified the control loop according to the difference between required and actual signal. The real time control is a necessary step to implement substructuring (see below). It may allow the possibility of precise control of more than one table to simulate differential effects.

Improved performance of high force capacity actuators, with maximum velocities up to 2m/s, with ±1m displacement, in order to be able to better represent recorded earthquake signals in the near-field.

Substructuring: PSD method already couples test results and the analysis of the tested structure. It can be generalized to more complex situations where part of the structure is completely analyzed and only a small part is tested. This has been successfully obtained in JRC-Ispra (CASCADE7, 2005), where a multispan bridge was studied: only the piles has been physically tested and a model of the bridge linked all the data. This must be generalized to continuous PSD and, next, to shaking table testing. One of the important questions to be solved is the delay between the measure and the force actuation.

Hybrid dynamic testing on shaking table, where part of the structure is modeled and its effect are simulated by actuators, as shown in Figure 16. This requires real-time control, substructuring, high-speed computation capacity of the modeled structure. The more intimate coupling between analysis and test is obviously made possible due to increase capabilities of computers. Taking into account the parallel increase in Networks capabilities, distributed testing is the next step to be considered. In this case, dynamic tests perfomed in different laboratories with different testing devices – shaking tables, centrifuge, reaction walls...- can be coupled, increasing significantly the testing capabilities.

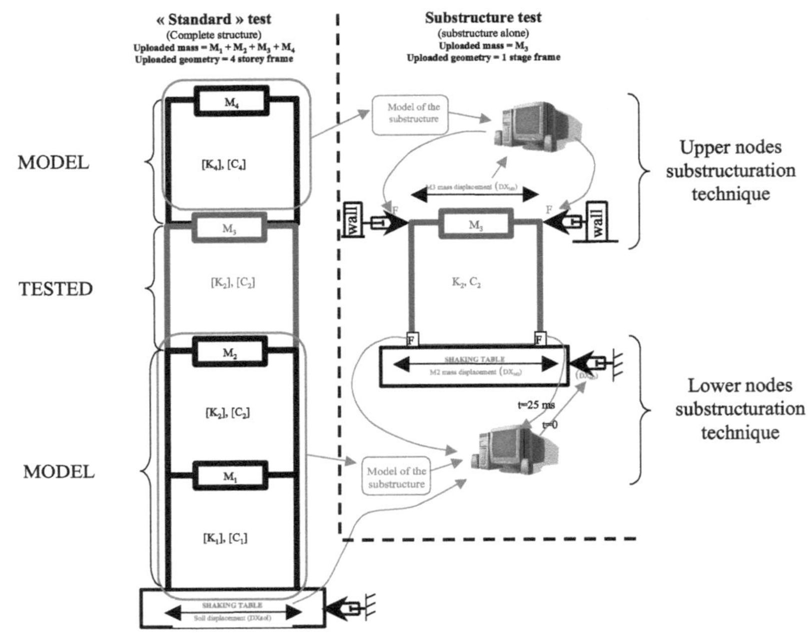

Figure 16. Substructuring and hybrid technique

These ingredients will constitute the basic bricks of the future advanced seismic test laboratory. This laboratory will include a network of scientific teams and facilities of different kind – reaction walls, shaking table and centrifuge, and dimensions. Shaking tables with high performances, expressed in terms of payload capacity and in term of actuators displacement and velocity, are a key element of the network. Facilities are linked by high capacity nets allowing exchange of data, remote participation to test, real time exchange of test data, calculation results for real time control of distributed test. Two examples, the first one in USA and the second in Europe are presented in Reinhorn et al., 2004 and Payen et al., 2006. The Figure 17 is a view of the European project mentioned in the latter reference; it has two "twin" moveable shaking tables and a reaction wall for hybrid testing.

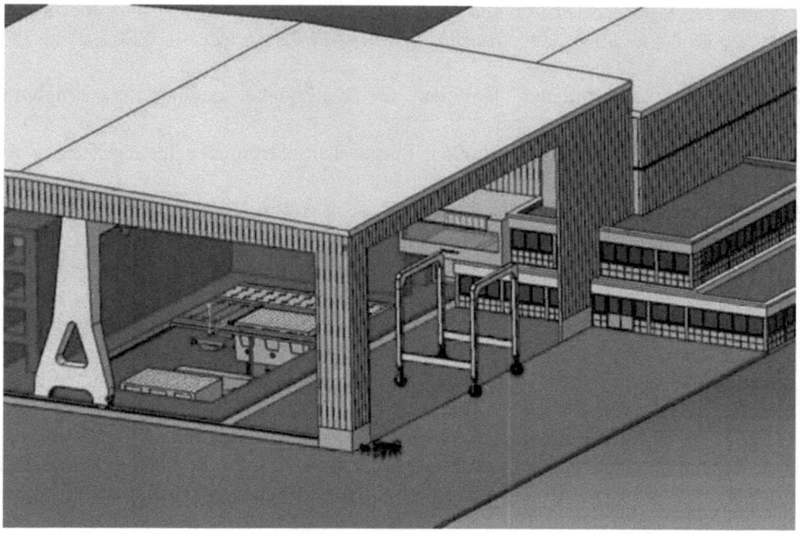

Figure 17. View of the European advanced seismic testing project

References

CASCADE2 (2005) Seismic qualification of high-voltage equipment. Report n°2. *A Research Training Network supported by European Commission* Published by LNEC (Lisbon)

CASCADE3 (2005) Dynamic qualification of electrical cabinets. Report n°3. *A Research Training Network supported by European Commission* Published by LNEC (Lisbon)

CASCADE6 (2005) Directory of european facilities for seismic and dynamic tests in support of industry. Report n°6. *A Research Training Network supported by European Commission* Published by LNEC (Lisbon)

CASCADE7 (2005) Recent advances and future needs in experimental earthquake engineering Report n°7 *A Research Training Network supported by European Commission* Published by LNEC (Lisbon)

Combescure, D. and Pegon, P., (1997). A-Operator splitting time integration technique for pseudodynamic testing. Error propagation analysis. *Soil Dynamics and Earthquake Engineering* 16. 427-443

Gajan, S., Kutter, B. L., Phalen, J. D., Hutchinson, T. C., and Martin, G. R. (2005). Centrifuge modeling of load-deformation behavior of rocking shallow foundations. *Soil Dynamics and Earthquake Engineering* 25 (2005) 773-783.

IEEE Std 693-1997. (1997) IEEE Recommended practice for seismic design of substations

IEEE Std 344-1987. (1987) IEEE Recommended practice for seismic qualification of class 1E equipment for nuclear power generating stations

Lee C-J. (2005). Centrifuge modelling of the behavior of caisson-type quay walls during earthquakes. *Soil Dynamics and Earthquake Engineering* 25 (2005) 117-131

Luong M.P. (1992). Centrifuge testing in *Recent advances in earthquake engineering.* V. Davidovici, Ed Ouest Editions

Negro, P., Pinto, A., V., Verzeletti, G. and Magonette, G. (1996). PsD test on a four-storey R/C building designed according to Eurocodes, *Journal of Structural Engineering.* ASCE, Vol 122, N°11, 1409-1417

OECD/NEA (2004). Experimental facilities for earthquake engineering simulation worldwide. NEA/CSNI/R(2004)10

Payen T., Quéval J.C., Sollogoub P., (2006). Large scale earthquake testing facility for vulnerability assessment. *1st ECEES*, Geneva.

Pegon, P. and Quéval, J.C., (2004). Structures sous séisme: expérimentation en laboratoires. In *Comportement dynamique des bétons et génie parasismique.* Lavoisier. Paris.

Reinhorn A.M. et al. (2004). The UB-NEES versatile high performance testing facility. *13th WCEE.* Vancouver

Semblat, J.F. and Luong, M.P. (1998). Wave propagation through soils in centrifuge testing. *Journal of Earthquake Engineering* Vol 2,N°1 (1998) 147-171

Shing, P., Vannan, M. and Cater, E. (1991) Implicit time integration for pseudodynamic tests. *EESD.* Vol 20; 551-556

Sieberg, A. (2005). *Experience and lessons on the origin, prevention and elimination of earthquake damages.* Reprint of 1943 version, with English translation. Sofia. Publishing House, LITSE.

Touboul F., Blay N., Sollogoub P. and Chapuliot S. 2006. Enhanced seismic criteria for piping. *Nuclear Engineering and Design* 236 (2006) 1-9

SPRINGER NATURE

GPSR Compliance

The European Union's (EU) General Product Safety Regulation (GPSR) is a set of rules that requires consumer products to be safe and our obligations to ensure this.

If you have any concerns about our products, you can contact us on ProductSafety@springernature.com

In case Publisher is established outside the EU, the EU authorized representative is:

Springer Nature Customer Service Center GmbH
Europaplatz 3
69115 Heidelberg, Germany

Kolibri 360 GmbH
Ferdinand-Jühlke-Straße 7
99095 Erfurt, Deutschland
produktsicherheit@kolibri360.de